XML in Technical Communication

Series profile

With a history dating back more than 50 years, the Institute of Scientific and Technical Communicators (ISTC) is the largest association representing information development professionals in the UK. It:

- Encourages professional education and supports standards
- Provides guidance about the value of using professional communicators
- Facilitates research, discussion and networking opportunities
- Liaises with other national and international technical communication associations.

The Institute's members create information that has an impact on people in virtually every sector of industry and society. They develop a wide range of information products, such as installation instructions for engineering equipment, operating and safety procedures for consumer products, and online user guides for computer software.

In their work, they face varied challenges that have an impact on their responsibilities, day-to-day tasks and careers. The ISTC provides opportunities for them to understand and overcome these challenges while fulfilling their core responsibility of communicating complex or important information in the most effective way for their readers.

One of the ways in which the Institute achieves this is to publish information in its monthly newsletter, *InfoPlus+*, and its quarterly journal, *Communicator*. It has now extended its publishing enterprise to launch this series of books, covering topics that have specific relevance to those working in technical communication.

If you would like to write for this series, please send your proposal to **books@istc.org.uk**.

For more information about the ISTC, visit **www.istc.org.uk**.

XML in Technical Communication

Charles Cowan

XML in Technical Communication

First published in Great Britain in 2008 by the Institute of Scientific and Technical Communicators (ISTC).

Second edition 2010

A CIP record for this book is available from the British Library

ISBN-10 09506459-82

ISBN-13 978-0-9506459-8-8

Typeset in Lucida Bright, Myriad Pro and Monospace 821 by WordMongers Ltd, **www.wordmongers.com**

Cover design by Sharp Jelly Creative, **www.sharpjelly.co.uk**

Institute of Scientific and Technical Communicators (ISTC)

Suite 111, Airport House, Purley Way, Croydon, CR0 0XZ

Telephone : +44 (0) 20 8253 4506

E-mail **istc@istc.org.uk**

Visit our website at **www.istc.org.uk**

Contents

Preface vii

1 XML Introduction 1
XML, the extensible markup language 1
General advantages of XML 4
XML in technical communication 5
XML and single sourcing 8

2 XML Essentials 11
XML markup 11
XML namespaces 19

3 Defining XML Languages 21
Schema languages 21
Document type definitions 22
XML Schema 39
Parsing and validating XML documents 52

4 XML Documentation Languages 55
DocBook 55
DITA – Darwin Information Typing Architecture 66
S1000D 85

5 Authoring with XML 125
Text editors 125
XML editors 125
XML and FrameMaker 128
XML and Word 133
XML and content management systems 136

6 Migrating to XML 137
What is involved in migration? 137
Migrating from HTML 139
Migrating from SGML 140
Migrating from FrameMaker 141

Migrating from Word 144
Migrating between XML languages 145

7 Transforming XML 147
Extensible Stylesheet Language 147
XSLT 148
XPath 150
XSL-FO 154

8 Using XML on the Web 157
XML and the web 157
Browser support for XML 158
XHTML 159
Cascading Style Sheets 161
XSLT stylesheets 163

9 XML and Localization 165
Overview 165
Advantages of XML for localization 166
XML in the localization process 172
Making XML documents easier to localize 176

10 Scalable Vector Graphics 177
Introduction 177
The advantages of Scalable Vector Graphics 178
Scalable Vector Graphics features 178
Some examples of SVG 180
Producing SVG documents 182
The future of SVG 185

11 XML and Content Management 187
Content management 187
Creation and authoring of content 188
Saving content with metadata 189
Storing content 190
Locating and retrieving content 191
Tracking changes to content 191
Publishing content 192
Controlling workflow 192

12 Summary 193
XML and technical communication tasks 193
The advantages of XML 196

A Glossary 201

B Bibliography 215

Index 221

Preface

In recent years Extensible Markup Language (XML) has had a major impact on software development and data interchange and has begun to have a similar effect in technical communication. XML provides a platform-independent, vendor-neutral source language for publishing documentation of all kinds and in any output format. XML solutions have already been adopted by many technical publication departments, while others are poised to make the move to XML. Authoring tools, publishing tools and content management systems are providing support for XML languages, in particular Darwin Information Typing Architecture (DITA), one of the XML languages designed specifically for technical documentation.

Increasingly, technical communicators therefore need to understand what XML has to offer and how it can be used in the various areas of technical communication. This book is an introduction to XML and its role in technical communication. It describes the basics of XML syntax and terminology, the use of XML in technical communication and the advantages it can bring to producing technical documentation of all types. It introduces the key XML languages that are used for technical documentation, either as documentation source formats or in transforming source documents to output formats.

Given the compact size of this book, none of the XML languages discussed in the book are described in detail. Rather, the book concentrates on the features and functionality of the various languages and technologies and their application to technical communication. The book therefore does not describe the details of how to implement XML solutions, but it does provide references to resources that you can use to do so.

A note on terminology may be helpful. *XML application*, *XML language* and *XML vocabulary* are all commonly used to refer to languages such as DocBook, MathML and even DITA. These terms are usually accepted as synonymous, though *XML vocabulary* is sometimes used with the narrower definition of '*A set of XML tags for a particular industry or business function.*' This book largely restricts itself to the term *XML language*.

The audience for this book

This book will be useful to anyone who needs an introduction to XML and its role in technical communication. For example, the book will be useful to:

- *Technical communicators*, in understanding the options for authoring in XML and the XML solutions and standards developed specifically for technical documentation, DocBook, DITA and S1000D.
- *Technical communicators* who need an introduction to the use and advantages of XML for internationalization.
- *Documentation managers*, in understanding the advantages that XML can provide in terms of increased productivity and cost reduction. The book will be helpful in answering questions about whether to use XML, which tools to use for authoring and processing XML and what is involved in migrating to the use of XML.

The book caters for technical communicators with various levels of markup language experience. The typical reader will probably have some experience of markup languages such as HTML, but the book can also be read by people with little or no XML experience. For readers of all experience levels, the book provides a comprehensive overview of the use of XML in technical communication, as well as references to more in-depth material.

Organization of this book

This book contains the following chapters and appendices:

- Chapter 1, *XML Introduction*. Provides an overview of XML, including its origins and relationship to other markup languages. The chapter discusses the general advantages of XML and its relevance to technical communication.
- Chapter 2, *XML Essentials*. Describes the different types of markup in an XML document, and introduces key XML terminology and concepts. The chapter introduces the syntax of XML by way of simple examples. If you are familiar with the basics of XML syntax, you might like to skim through or skip this chapter.

Chapter 3, *Defining XML Languages.* Discusses, with examples, the two main schema languages for defining the markup available in XML languages: Document Type Definitions (DTD) and the XML Schema language.

The chapter describes enough about DTDs to allow you to understand and write them if you need to do so. As technical communicators, you might be involved in writing or modifying DTDs to extend the capabilities of the XML language that you are using.

If your role does not require you to know about how XML languages are defined, you might like to skip this chapter on your first reading of this book and return to it later to gain a deeper understanding of XML.

Chapter 4, *XML Documentation Languages.* Introduces and compares the main XML languages and standards designed specifically for producing technical documentation: DocBook and Darwin Information Typing Architecture (DITA) and the S1000D specification. The chapter describes the main features of these languages and the tools that support them.

Chapter 5, *Authoring with XML.* Describes the options available for authoring XML documents and the features that you should expect in an XML editor. The chapter also discusses support for XML in the most commonly used publishing tools such as FrameMaker and Word.

Chapter 6, *Migrating to XML.* Discusses migration to XML as the source format for your documentation. The chapter discusses general migration considerations, as well as providing information specific to migration from HTML, SGML, Adobe FrameMaker and Microsoft Word.

Chapter 7, *Transforming XML.* Describes the technologies used to transform XML documents to other formats, including HTML, other XML languages and PDF. The chapter covers the two components of Extensible Style Language (XSL): XSL Transformation (XSLT) and XSL Formatting Objects (XSL-FO), as well as the XPath language that is used in conjunction with XSLT.

Chapter 8, *Using XML on the Web.* Describes how XML is used in web pages, either directly as XHTML, or indirectly as XML documents with Cascading Style Sheets (CSS) or XSLT stylesheets that specify how to render the XML as HTML. The chapter discusses the extent to which web browsers support XHTML, XSLT and CSS.

Chapter 9, *XML and Localization.* Describes the advantages of XML for localization of documentation and the cost savings it can provide. The chapter summarizes the ways in which XML is used in localization, and introduces the XML languages that have been developed specifically for use in localization.

- Chapter 10, *Scalable Vector Graphics.* Discusses SVG, the XML standard for describing vector graphics. The chapter explains the advantages of using SVG in technical documentation and summarizes the support available for SVG in publishing tools, graphics tools and web browsers.
- Chapter 11, *XML and Content Management.* Discusses the advantages of XML in content management systems (CMS). The chapter describes how XML facilitates storage, location, retrieval and reuse of content within a CMS.
- Chapter 12, *Summary.* Summarizes the advantages that XML provides for technical communicators and how the features and facilities of XML provide those advantages. The chapter also provides a summary table showing how common technical communication tasks are facilitated by XML.

A glossary and bibliography complete the book, providing a definition of terms used and a list of key XML-related web resources, books and articles.

1

XML Introduction

This chapter:

- Provides an introduction to XML, describing its origins and its development
- Summarizes the features of XML, describing its general advantages
- Summarizes how XML is used in technical communication
- Describes how XML is ideally suited for single-source publishing.

XML, the extensible markup language

The story of XML really begins with Standard Generalized Markup Language (SGML). SGML was developed in the 1970s to provide a standard format to support sharing of documents between different programs. The problem was that you could not easily use documents written using one program and saved in a specific format with another program, particularly one running on a different operating system. SGML became an ISO standard (ISO 8879:1986) in 1986.

SGML itself is actually a *metalanguage*, a language for defining markup languages. SGML specifies the rules and general syntax for such markup languages, which are also called *applications*. For example, the markup language DocBook Version 3.1 is an SGML application, as is the markup language IBMIDDOC, an SGML application used by IBM as a source format for technical publications.

There are various types of markup language, but essentially a markup language is a set of codes or tags that surround content and describe what that content is, or in some cases what it should look like when displayed. In SGML (and XML) applications, therefore, tags surround the document content, and the tags may have attributes that further qualify the context of the content. For example, the `<book>` tags in the following code indicate that the content of the document describes a book, and the author, title and subject tags indicate the meaning of the content. The `audience` attribute of the `<subject>` tag provides further information about the content.

```
<book>
     <author>Peter Wrightwell</author>
     <title>XML in Technical Publication</title>
     <subject audience="technical">XML</subject>
</book>
```

The tags surrounding the content and the content itself constitute an element. For example, `<author>Peter Wrightwell</author>` is an element, and `<author>` and `</author>` are *start tags* and *end tags* respectively. People tend to use the terms *tag* and *element* interchangeably, but strictly speaking, *tag* refers only to the start tags and end tags in elements.

By authoring a text document using a markup language and keeping presentation information separate, you enable both sharing of documents that are readable by computers and humans and single-source publishing. For example, if you use the DocBook language as the source format for documents, presentation information in different stylesheets can be applied to the documents according to the required output formats for publishing.

SGML applications were adopted and are still used as a publishing source format by a number of industries and organizations, but SGML applications, with their large number of elements, were considered too complicated and unsuitable for small-scale general-purpose use – many preferred the relative ease of use and WYSIWYG environment of software such as Word and FrameMaker.

When the World Wide Web appeared, one SGML application, Hypertext Markup Language (HTML), was adopted as the source markup language for web pages. HTML is relatively easy to use, but there are reasons why it is not suitable as a standard documentation format. Different versions of HTML were developed as different web browsers vied for domination: for example, some elements supported by Internet Explorer were not supported by Netscape, and vice versa. HTML is a markup language that defines how content is presented, but it does not describe the structure or meaning of the content.

In the 1990s the World Wide Web Consortium (W3C) therefore started an initiative to introduce a less complicated version of SGML that could be used on the web and elsewhere, which would overcome the incompatibility problems of HTML, and which would be easy to author and process. This resulted in Extensible Markup Language (XML), with XML 1.0 being published in 1998 and XML 1.1 in 2004 (see Figure 1.1). XML has quite a few differences from SGML, the most obvious of which is probably its case sensitivity. In SGML, for example, the tags `<BOOK>` and `</book>` are acceptable, while in XML the tags must have the same case, for example `<book>` and `</book>`. For a detailed comparison of XML and SGML, see `www.w3.org/TR/NOTE-sgml-xml.html`.

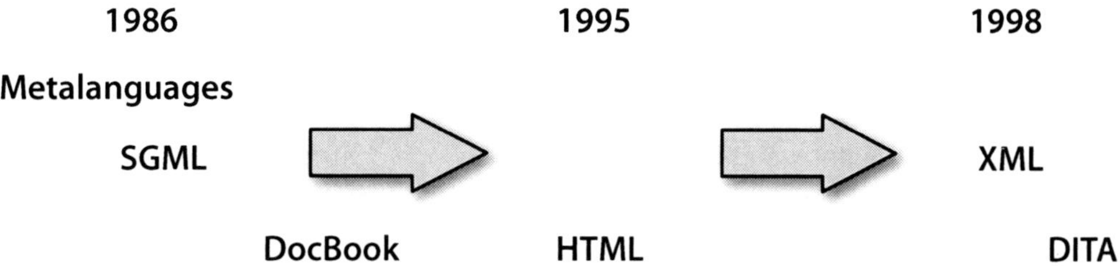

Figure 1.1 XML and markup language timeline

Since its inception XML has achieved widespread use as a source markup language for technical documentation, and can also be used on the web. However, it is also used for many data-centric purposes in addition to the publishing applications for which SGML was used. XML has attained a central role in programming and software development: it provides a standard data interchange format, a format for extracting data from databases and a versatile content storage format.

The W3C has been responsible for the development of XML, key *XML languages* and supporting technologies, as well as standards for the World Wide Web. The W3C consists of member organizations and individuals, and is an international standards organization that follows a ratification process in the development of its standards. A W3C Recommendation is the final stage in this process, and is the equivalent of a published standard in other industries. Apart from the XML standard, there are Recommendations for:

- XSL – the family of recommendations for defining XML document transformation and presentation

- XLink – the language for defining links between and within XML documents and other documents
- XHTML – the XML version of HTML
- XML Schema – an XML language for defining the structure, content and semantics of XML documents.

Many other XML-related technologies also exist.

General advantages of XML

Like SGML, XML is a metalanguage. There is no fixed set of elements in XML. You can define your own XML languages with whatever elements you like and add further elements to any XML language. This is what is meant by *extensible* in the XML name. XML languages have been developed as documentation source formats (for example, the XML version of DocBook and DITA), and for many other purposes. The following are just some examples of the many available XML languages:

- *Mathematical Markup Language* (MathML), for embedding mathematical equations in web pages and other documents.
- *Wireless Markup Language* (WML), for storing information for display on mobile devices.
- *Chemical Markup Language* (CML), for managing molecular information.
- *Really Simple Syndication* (RSS), for publishing frequently updated web content such as blog entries, news feeds, or podcasts.

These special-purpose languages are all based on XML syntax, but they have elements appropriate to their purpose. For example, MathML has elements for describing mathematical notations, while CML has elements such as `<atom>` and `<molecule>` for representing molecular information. The advantage of XML's extensibility is that you can customize and adapt an XML language to your needs.

Although you can define XML elements with any name you like, one of the advantages of XML is the ability to create a hierarchical structure of elements with meaningful names, which is called *semantic tagging.* The attributes of the elements can provide further information, *metadata,* about the content. This makes it easier for humans to read an XML document and understand the meaning of the content, but also means that software can search and index XML documents in a meaningful way. The hierarchy of elements in an XML document adds structure to the content: by validating the document against a description of the markup allowed for the XML language (a *schema document*), you can ensure that all XML documents in a particular language have a consistent structure. The ways in which you can define the markup allowed for an XML language are described in Chapter 3, *Defining XML Languages.*

The semantic nature of XML also means that XML is a foundation technology in the Semantic Web, a vision of a future World Wide Web in which you can search for information and resources according to their meaning. Although the Semantic Web does not yet really exist, it is likely that Resource Description Framework (RDF), an XML language designed as a standard for describing the meaning of resources, will play a central role. XML is also used in another component of the Semantic Web, *ontologies*, which are formal descriptions of resources and their relationships.

As XML is effectively an open, non-proprietary standard and universal data format, it allows for portability of information across operating systems and manufacturers. XML documents are easily transferred, imported into and exported from software regardless of the operating system on which it is running - Windows, UNIX, Macintosh or other platforms. You can produce an XML document from a program under Windows and use the same document in a program running on a Macintosh or on UNIX.

As XML is a text format, it is easily processed and transformed into other formats by software. A wealth of tools has been developed for processing XML, and many books about XML and its supporting technologies are available.

XML in technical communication

Figure 1.2 illustrates how XML is deployed in the development of technical documentation, from its use in authoring tools, through the use of conversion tools, to the output of technical documents in various formats.

Various XML languages are available as documentation source formats. DocBook is well established and was originally intended for hardware and software technical documentation, but it can be used for any other sort of documentation. Darwin Information Typing Architecture (DITA) is an XML-based language and architecture for authoring, producing and delivering technical information that, like DocBook, has become an

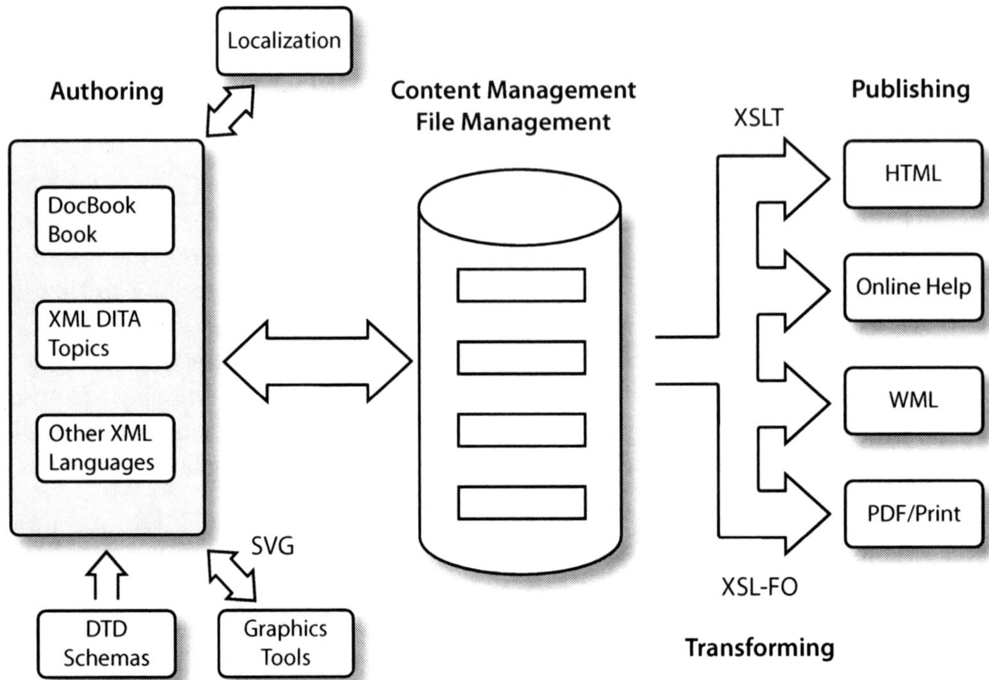

Figure 1.2 XML authoring and publishing

OASIS (Organization for the Advancement of Structured Information Standards) standard. Both languages are well supported by processing tools and have active user communities: they are discussed in detail in Chapter 4, *XML Documentation Languages.* Although DocBook and DITA are arguably the leading XML languages used in technical documentation, the international specification S1000D can be implemented in XML and is widely used for equipment maintenance and operations publications.

A growing number of tools are available for authoring XML. Some, for example XMetal and XMLSpy, support XML as their native format and provide WYSIWYG and structure views in addition to code views of XML documents. Others, such as FrameMaker, allow import and export of documents to various XML formats. For example, structured FrameMaker as supplied supports DocBook, DITA, S1000D and XHTML. Authors can import XML to take advantage of the publishing facilities of FrameMaker, for example to produce high-quality PDF, and export to XML formats for output or for further processing. When authoring XML, it is essential that the XML source files are validated against the appropriate Document Type Definition (DTD) or schema document. DTD and XML

Schema and their role in defining the markup allowed in an XML language are described in Chapter 3, *Defining XML Languages.*

Many other publishing tools support XML. Some authoring tools are essentially *content management* systems that provide XML authoring as well as the management, storage and version control facilities expected of a content management system. PTC Arbortext is an example of such a tool. Other tools that support XML authoring might integrate with CMS tools that do not in themselves support XML authoring. In either case, authors can check files out and back into a repository when they need to work on them. Options available for authoring in XML are discussed in Chapter 5, *Authoring with XML.*

Although there are some XML languages that deal with presentation, for example Extensible HyperText Markup Language (XHTML), source documents marked up in XML languages such as DocBook do not contain information about how the content is presented when printed or displayed on the web. You can supply the presentation information when the XML documents are converted to output formats by applying an extensible style language transformation (XSLT) stylesheet. XSLT is itself an XML language that you can use to transform XML documents to HTML, other XML languages or indeed any text-based documentation format. XSLT can also transform XML documents to the XSL-FO format, an XML page description language, which is then converted through the use of suitable tools to PDF or other formats suitable for printing. XSLT and XSL-FO are discussed in Chapter 7, *Transforming XML.*

XML was originally intended as a format for the web, and although it has been used for many other purposes, it is also used in web pages. Some web browsers support XML natively: that is, they can take XML documents associated with CSS (Cascading Style Sheet) or XSLT stylesheets and produce HTML files on the fly. Another possibility is XHTML, an XML language that is effectively HTML reformulated as XML. It has similar elements to HTML and is also designed for web pages, but follows the syntax rules of XML. XHTML has several advantages over HTML: for example, it provides a standard set of elements, and it can exploit XML processing tools. The use of XML on the web is discussed in Chapter 8, *Using XML on the Web.*

Technical publications must often be translated to other languages to support software products that are produced in localized versions. XML languages provide standard formats that are used easily by translators and exchanged between translation products. XML's support of Unicode is a huge advantage for localization, and means that XML documents can represent text in any of the world's languages. XML has various other advantages for localization and many features that facilitate translation. These are discussed in Chapter 9, *XML and Localization.*

Scalable Vector Graphics (SVG) is an XML language that is set to be the standard technology for displaying high-quality graphics on the web and elsewhere. SVG allows web graphics to be interactive and animated, while their textual content is easily searched, accessible and displayable in multiple languages. You can include SVG markup within XML documents and you can export graphics as SVG documents from graphics tools such as CorelDraw and Adobe Illustrator. SVG is discussed in Chapter 10, *Scalable Vector Graphics.*

Effective content management is essential for storing and managing an organization's data, and here again XML can play an important role. XML has many advantages as a storage format for content management systems: its structured and semantic nature allows granular management and categorization of content within a CMS. XML can facilitate searching and retrieval of information, tracking changes and controlling workflow. XML content management is discussed in Chapter 11, *XML and Content Management.*

XML and single sourcing

Although it is not the only solution for *single source publishing*, XML has a number of features that make it ideal. Single source publishing (also known just as 'single-sourcing') allows the following types of reuse:

- *Reassembly* - using modules of content in a number of documents. Separate modules are maintained in one place and assembled into different documents as and when required, possibly at the time of publication.
- *Repurposing* - delivering documentation in multiple output formats from a single source format. For example, using the same content in online help, a printed PDF document and in web pages.

These are discussed below.

Reassembly

With XML, complete pages or smaller units of content can be treated as separate modules. For example, you can create modules that represent sections, paragraphs or procedures. You can then reassemble these modules for whatever publishing purpose or audience you require. For example, you might have a module containing the markup elements for an installation procedure that is used in a number of manuals, or a topic that is relevant to a number of publications.

As another example, you might apply conditional processing to assemble from your repository only the modules of content required for a particular audience. In this way, you eliminate the duplication of content in the different publications, and thereby reduce maintenance and translation costs, improve consistency and minimize errors.

Repurposing

Repurposing is facilitated by XML's separation of content from format, by keeping formatting information in a separate stylesheet. You can apply a stylesheet to provide consistent formatting information to multiple XML documents, and you can apply different stylesheets to a single XML document to produce different output formats.

XML is also well suited to repurposing, because XML documents are easily transformed to output formats such as HTML, PDF, WML and so on. Repurposing saves time and money because:

- You only have to maintain content in one place
- You can maintain presentation information in a small number of places
- You do not have to expend effort in producing and formatting the same content in multiple formats.

2

XML Essentials

This chapter:

- Describes the markup, the syntactic components, that constitute an XML document
- Explains key XML terminology, in particular the terms *well formed* and *valid*
- Introduces the important concept of XML Namespaces.

If you already know the basics of XML syntax, you may like to skip this chapter.

XML markup

The XML document shown in Figure 2.1 is an example in the DocBook language that represents a book consisting of two extremely short chapters. The example is certainly contrived, but it illustrates all the types of markup that can appear in an XML document. The following sections describe each of these markup types, using the line numbers in the figure for reference.

An XML document such as this example does not specify how the information in the document is presented: that is done using a stylesheet when the document is transformed to an output format such as HTML or PDF. Figure 2.2 illustrates how the example XML document might appear in a web browser when transformed to HTML. In this case, the transformation process used the `docbook.xsl` stylesheet, as shown in line 6 in Figure 2.1.

```
<?xml version="1.0" encoding="UTF-8"?>
<!DOCTYPE book PUBLIC "-//OASIS//DTD DocBook XML V4.2//EN"
          "http://www.oasis-open.org/docbook/xml/4.2/docbookx.dtd" [
  <!ENTITY booktitle "XML Examples">
]>
<?xml-stylesheet type="text/xsl" href="docbook.xsl"?>
<book>
  <bookinfo>
    <title>&booktitle;</title>
    <author>
      <firstname>Charles</firstname>
      <surname>Cowan</surname>
    </author>
  </bookinfo>
  <chapter id="ch01">
    <title>The First Chapter</title>
    <para> This chapter only links to the next chapter.
   For more information, see <xref linkend="ch02"/>.
    </para>
  </chapter>
  <chapter id="ch02">
    <title>The Second Chapter</title>
    <!-- Three ways of tagging markup characters -->
    <para>In DocBook, the &lt;xref&gt; element is an
    <glossterm>empty element</glossterm>.
    It is not the same as the <![CDATA[  <Xref> ]]> element, or the
    &#60;XREF&#62; element, because XML is case sensitive!
    </para>
  </chapter>
</book>
```

Figure 2.1 An example XML document in the DocBook language

XML Examples

Charles Cowan

Table of Contents

1. The First Chapter
2. The Second Chapter

Chapter 1. The First Chapter

This chapter only links to the next chapter. For more information, see Chapter 2.

Chapter 2. The Second Chapter

In DocBook, the <xref> element is an *empty element*. It is not the same as the <Xref> element, or the <XREF> element, because XML is case sensitive!

Figure 2.2 The example document transformed to HTML and displayed in a browser

XML declarations (line 1)

The first line of an XML document can contain the *XML declaration*, which specifies the version of XML and the character encoding used in the document. The XML declaration is optional: if it is not included in the document, XML Version 1.0 and the default encoding of UTF-8 (8-bit Unicode Transformation Format) are assumed.

UTF-8 is a character encoding for Unicode, the standard for the representation of characters of all languages. This means that the markup and content of XML documents can contain text from practically all of the world's languages. For more information about XML encoding, see Chapter 9, *XML and Localization*.

Document type declarations (lines 2–5)

The XML declaration can be followed by a further declaration, the *document type declaration*. This declaration either contains or refers to a Document Type Definition (DTD). A DTD defines the *elements*, *attributes* and other markup that you can use in an XML document. As shown in Figure 2.1, the document type declaration can both refer to a DTD (`docbookx.dtd` in the example) and contain additional declarations (an *internal subset*), such as the entity definition in line 4. For information about the use of DTDs, see Chapter 3, *Defining XML Languages*.

Not all XML documents have a DTD: there are other mechanisms for defining XML languages, and, for example, there are uses of XML in data processing that do not require documents to have a DTD. In such cases, the XML declaration can contain the attribute `standalone = "yes"` to specify that a document does not have a DTD.

Elements (line 7 onwards)

An XML document contains a hierarchy of elements that represents the structure of the information in the document. The first element in the document is the *root element* (also called the *document element*) and contains all the other elements in the document. In the example in Figure 2.1, the root element is `<book>`. An element contained by another element is called a *child element* and the containing element is known as the *parent element*.

An XML element usually consists of a *start tag* and an *end tag*, which enclose the *content* of the element. For example, the `<firstname>Charles</firstname>` element consists of the start tag `<firstname>`, the content `Charles`, and the end tag `</firstname>`. There are however, elements that are termed *empty elements* because they have no content. Empty elements have only a single tag with the syntax `<tag/>`. An example of an empty element is DocBook's `<xref>` element, `<xref linkend = "ch02"/>` in line 18 of the example in Figure 2.1.

The content of an element can consist of:

- Character data alone: see for example the `<title>` element in line 9 of Figure 2.1
- Zero or more elements: see for example, the `<author>` element in line 10
- A mixture of character data and elements, called *mixed content*: see for example the `<para>` element in line 17.

In an XML document you can and should lay out the elements with indentation to show their hierarchy and to make the document easier to read. In XML, whitespace characters (tabs, linefeeds, carriage returns and spaces) between elements are usually preserved by processing software, which is not the case for HTML or SGML.

Attributes (line 18)

Elements can have attributes, for example, the element `<xref linkend = "ch02"/>` has the attribute `linkend` which has the value `ch02`. Attribute values must be enclosed in single or double straight quotes.

Attributes provide further information (*metadata*) about an element or its content. For example, the `linkend` attribute of `<xref>` specifies the target of a cross-reference. Of course, an XML language could achieve the same result by using child elements rather than attributes, and there is a great deal of debate about the merits of each approach. There are no hard and fast rules, but generally where structure is important you are likely to find that child elements are used rather than attributes.

Attributes are used to provide information to processing software as well as information that is human-readable. Typical uses in technical documentation include:

- *Conditional processing.* For example, attributes can identify document content that applies to a particular version of a product, or to a particular operating system such as Windows or UNIX. An example is DocBook's `vendor` attribute, which indicates the computer vendor to which the element content applies.
- *Repurposing.* For example, an attribute can specify whether the content of an element is applicable for particular output formats such as HTML or PDF.
- *Cross-references.* Attributes provide identifiers for elements and specify targets for cross-references. For example, in DocBook the `id` attribute is used as the identifier for elements, and the `linkend`

attribute of the <xref> element specifies the identifier to which the cross-reference links.

- *Information about graphics.* For example, attributes can specify the width, height, alignment and format of graphics.
- *Translation.* An attribute can specify whether or not the content of an element (and its children) should be translated.
- *Search information.* Attributes can provide key words that are used by search engines.

There are other uses of attributes: for example, they can provide tracking information during information development, and they can provide metadata about publications such as author and publisher details.

You can use some special attributes on any element:

- The attribute `xml:lang` identifies the language of the content of the element (see Chapter 9, *XML and Localization*).
- The attribute `xml:space` specifies whether whitespace characters within the content of an element are preserved by processing software. The possible attribute values are `preserve` and `discard`, which is the default value. This is useful for preserving text formatting within element content, for example in examples of software code.

There are strict rules for the names of elements and attributes. *XML names* can contain any alphanumeric characters, both English characters and characters from other languages, and also the underscore (_), hyphen (-) or period (.) characters. They cannot contain any other punctuation characters such as quotation marks, dollar signs, slashes or percent symbols, or whitespace characters of any kind. Names cannot begin with a numeric character, hyphen or period: they must begin with an alphabetic character or an underscore. Table 2.1 illustrates some valid and invalid element names:

Table 2.1 Valid and invalid element names

Valid names	***Invalid names***	***Reason***
<_firstname>	<$firstname>	The $ character is not allowed
<to_be_sure> <first-name>	<2BSURE> <.less>	Names cannot begin with numeric characters or periods
<μ>	<first name>	Names cannot contain spaces

XML is case sensitive, so the elements `<title>` and `<TITLE>` are not equivalent, as they would be in HTML.

Character references (line 27)

Character references are used to insert single characters, such as special characters and symbols, that you cannot otherwise enter into an XML document. Examples include Chinese characters if you are using European-language software settings, or accented characters that are not available in your text editor.

Character references consist of an ampersand, the hash character, a number and a semicolon. For example, the character reference `Ç` is used for the capital C character with a cedilla (Ç). The number can be a decimal or hexadecimal number (in which case, an `x` character precedes the number) and corresponds to a Unicode character value.

Fortunately you do not have to remember all the numbers for character references, as you can use entity references for single characters, as described in the following section, and XML and other editors provide easy methods for inserting character references.

Entity references (lines 9, 24)

In the example document `&booktitle;` is an example of an *entity reference*. When an XML document is processed, entity references such as this are replaced by the value from the corresponding *entity definition* in the DTD. In the example document, for example, the entity reference `&booktitle;` will be replaced by the text `XML Examples` from the entity declared in line 4:

```
<!ENTITY booktitle "XML Examples">
```

Entity references consist of the name of the entity preceded by an ampersand character and followed by a semicolon, as in `&booktitle;` in the example document in Figure 2.1.

You use entity references to provide replacement text for variables. They are invaluable as a means of reuse in technical documentation, ensuring easy maintenance of documentation when product names and terminology change, as well as consistency of spelling. The replacement text can consist of single words or large sections of boilerplate text, and you can even include whole files using this mechanism. You can also use entity references instead of character references. For example, you could use the entity reference `Ç` instead of the character reference `Ç` for the Ç character.

Entities must be declared in a DTD or document type declaration, so, for example, to use the `Ç` entity reference, you can use the following entity declaration:

```
<!ENTITY Ccedil "&#199">
```

Entity declarations for natural language, mathematical, and other special symbols are available with XML languages such as DocBook; in fact these entity declarations are maintained as a standard set by the W3C (see `www.w3.org/2003/entities`). However, you can use five predefined entity references that do not require any entity declaration. They are:

Entity reference	*Character*
`&`	The ampersand (&)
`<`	The less than character (<)
`>`	The greater than character (>)
`"`	The single straight quote (')
`'`	The apostrophe (‘)

These entities allow you to use the characters that would be confused with markup, for example the '<' character, in the content of elements and in the values of attributes (see line 24 in Figure 2.1).

Processing instructions (line 6)

An XML document can contain processing instructions that are used to pass information to software that processes the XML document. Processing instructions have the format:

```
<? target data ?>
```

where *target* is a label or else identifies the processing software, and *data* is information that the processing software uses.

In technical documentation, processing instructions are used mainly for identifying a stylesheet to be used in rendering the XML document. For example, the processing instruction in line 6 of the document in Figure 2.1 informs the processing software that the `docbook.xsl` stylesheet must be used in transforming the XML document to HTML:

```
<?xml-stylesheet type="text/xsl" href="docbook.xsl"?>
```

As another example, the following processing instruction tells a web browser to apply the `book.css` stylesheet before displaying the XML document:

```
<?xml-stylesheet href="book.css" type="text/css"?>
```

XML editors also use processing instructions, for example to provide bookmark information and other information used in editing sessions.

CDATA sections (line 26)

Another possible constituent of XML documents is the CDATA section, which has this syntax:

```
<! [CDATA [ character data ]]>
```

CDATA sections are used to identify data that processing software should not interpret as markup, and are used to include sections of XML, HTML and other code within XML documents. They are also useful for the convenient input of text that would otherwise contain many entity references.

Comments (line 23)

An XML document can contain comments enclosed within the `<!--` and `-->` characters, the same syntax as in HTML. Comments are useful for providing document history and other information for authors of XML documents. You can place comments anywhere within an XML document except within tags or processing instructions.

XML processing software does not regard comments as XML markup, so comments are ignored, for example when XML is transformed to HTML.

Well-formed and valid XML documents

All XML documents must be *well formed,* that is, conform to the rules of XML syntax. There are many rules to which well-formed documents must conform, and some of them are quite obscure. The full list of rules is given in the W3C XML Specification (`www.w3.org/TR/REC-xml`), but the main rules are shown in Table 2.2.

As you can see, the markup rules required for XML are much stricter than those of HTML. With HTML, browsers allow markup that is not allowed in XML, such as elements that do not nest properly and elements that do not have end tags.

An XML document can also be *valid,* which means that it conforms to the rules defined in a DTD or other *schema document.* For example, a DTD defines the elements that can appear in an XML document of the type

Rule	*Well formed*	*Not well formed*
There must be only one root element.	<?xml version='1.0'?> <root> ... </root>	<?xml version='1.0'?> <root> ... </root> <root2> … </root2>
For each start tag there must be a matching end tag (remember that XML is case sensitive).	<start> </start>	<start> </START> <start> text <elem> text </elem>
Elements must not overlap, that is, they must nest properly.	<element1> <element2> </element2> </element1>	<element1> <element2> </element1> </element2>
All attribute values must be enclosed in single or double straight quotes, not curly quotes.	<elem attribute = "value"/> <elem attribute = 'value'/>	<elem attribute = “value”/>
An element cannot have two attributes of the same name.	<elem attr1 = "yes" attr2 = "yes"/>	<elem attr1 = "yes" attr1 = "no"/>
Comments and processing instructions cannot appear inside tags.	<elem> Text <!--comment --> more text </elem>	<elem <!--comment -->> Text </elem>
The < and & characters must not appear as such within character data.	<elem> 4 < 6 & 5 < 3</elem>	<elem> 4 < 6 & 5 < 3 </elem>

defined by the DTD. If an XML document contains any other elements, or if it contains attributes not declared in the DTD, then it is said not to be valid. Validating parser software is used to check the validity of XML documents, as described in Chapter 3, *Defining XML Languages.*

Sometimes it is necessary to use markup from different XML languages within a single XML document. For example, you might want to include Scalable Vector Graphics (SVG) markup within your XML document. In such cases, the different XML languages might have elements of the same name: for example, the SVG language has a `<title>` element, and it is quite likely that your XML source language also has a `<title>` element (as DocBook does).

Luckily, there is a mechanism called XML Namespaces that overcomes this problem. A *namespace* is a container that provides context for the markup. By associating an element or attribute with a namespace, you can distinguish elements of the same name.

In an XML document, you identify namespaces with the `xmlns` attribute, which has the syntax:

```
xmlns:prefix="URI"
```

where the *prefix* is used to bind elements to a Uniform Resource Indicator (URI), which is similar to a URL and which uniquely identifies the namespace. A namespace is just an identifier: there does not need to be anything at the URI defined for the namespace. The important point is that the URI mechanism ensures that the namespace is unique.

You can use the `xmlns` attribute on the root element or any of its descendents. For example, the following markup identifies the URI for the SVG namespace:

```
<root xmlns:svg="http://www.w3.org/2000/svg">
```

Within the XML document, you can then associate elements and attributes with the namespace by qualifying them with the prefix `svg`. For example, for the SVG `<title>` element:

```
<svg:title>
```

You can define multiple namespaces within an XML document: markup belongs to a particular namespace until the next namespace declaration in the document. You can also define a default namespace using slightly different syntax for the `xmlns` attribute:

```
xmlns="URI"
```

For example:

```
<svg xmlns="http://www.w3.org/2000/svg">
```

You do not have to qualify elements or attributes that belong to the default namespace. If an XML document uses elements with prefixes and has a DTD, those elements must be declared in the DTD. For more information about namespaces and their relationship to DTDs and other schema documents, see Chapter 3, *Defining XML Languages.*

3

Defining XML Languages

This chapter describes the two main ways of defining an XML language:

- Using Document Type Definitions (DTD)
- Using the XML Schema language.

Schema languages

An XML schema document defines the elements, attributes and other markup that can appear in a document marked up in a particular XML language. In other words, it specifies both the type of information that can appear in the document and its structure. An XML language is also known as an *XML vocabulary* or *document type.*

Schema documents are written in an XML schema language. The two main languages in use are Document Type Definition (DTD) and the W3C XML Schema language. There are other schema languages, for example RELAX NG and Schematron, but DTD and XML Schema have greater significance to technical communication.

When you validate your XML documents against an XML schema document, you ensure that all documents of a particular type have a consistent structure. Furthermore, when you use an XML editing tool in conjunction with a schema document, you can ensure that all required elements are present and in the correct hierarchy in your XML documents as you author them.

Document type

The acronym 'DTD' is used both to refer to the language and to documents written in the DTD language. A DTD can contain the following types of statement:

- *Element declaration.* The name of an element and what it can contain.
- *Attribute declaration.* The name of an attribute and its type, as well as any default value for the attribute.
- *Entity declaration.* A definition of an entity: either a *general entity* for use in an XML document, or a *parameter entity* for use within the DTD itself.
- *Entity reference.* A reference to a parameter entity declared in the DTD that enables reuse of element and attribute declarations within the DTD.
- *Notation declaration.* Information related to non-XML content associated with an XML document.
- *Comments.* Information to help you understand the DTD.

The following sections describe how you declare each type of markup in a DTD.

You declare elements using the `<!ELEMENT` keyword with the following format:

```
<!ELEMENT name content_specification>
```

The *name* is the name of the element and must be a valid XML name. The *content specification,* also known as the *content model,* defines what the element can contain in terms of character data (text) and child elements. There are five types of content specification:

- `PCDATA`. The element contains character data only and no child elements.
- *Empty elements.* The element contains neither character data nor child elements, but can have attributes.
- *Elements only.* The element contains child elements only.
- *Mixed content.* The element contains child elements and character data.

- *Any content.* The element contains any combination of child elements and character data.

The following sections describe how to declare elements of each type.

Note that there is no explicit declaration of a root element in a DTD: that is done in the document type declaration of an XML document. A root element must not be in the content specification of any other element in the DTD.

Declaring PCDATA elements

To declare an element that can contain text only and no child elements, you use `(#PCDATA)` in the content specification. 'PCDATA' stands for 'parsed character data', which includes simple text and can include character and entity references, (as well as `CDATA` sections, processing instructions and comments). The essential point about parsed character data is that it cannot contain elements. For example:

```
<!ELEMENT product_name (#PCDATA)>
```

declares an element `<product_name>`, which might contain the following text in an XML document:

```
<product_name> This is text only with the &prodname;
</product_name>
```

Declaring empty elements

To declare an empty element you use `EMPTY` in the content specification. For example:

```
<!ELEMENT image EMPTY>
```

declares an empty element `<image>`, which might appear in an XML document as follows:

```
<image id = "1234"/>
```

Empty elements are used, for example, for elements that represent cross-references, or for graphic images.

Declaring elements containing only child elements

To declare an element that can contain one or more child elements only, you specify the names of the child elements within parentheses. If the child elements follow a sequence, you define the order using the comma (`,`) character. For example:

```
<!ELEMENT name (child1, child2, child3)>
```

declares that the <name> element must contain the <child1>, <child2>, and <child3> elements in that order. The <name> element might then appear in an XML document as follows:

```
<name>
    <child1>Peter</child>
    <child2>Paul</child>
    <child3>Mary</child>
</name>
```

To define an optional set of elements, you use the pipe (|) character. For example:

```
<!ELEMENT name (child1 | child2 | child3)>
```

declares that the <name> element must contain one of the <child1>, <child2> or <child3> elements.

You define the *frequency* of child elements, that is, the number of times they can appear within the parent element, using one of the *occurrence indicator* characters:

- *Question mark (?).* Zero or one of the child elements must appear: in other words, the element is optional but can only appear once.
- *Asterisk (*).* Zero or more of the child elements must appear: in other words, the element is optional and can appear multiple times.
- *Plus sign (+).* One or more of the child elements must appear.

For example:

```
<!ELEMENT name (child1?, child2*, child3+)>
```

declares that the <name> element can contain one <child1> element, one or more <child2> elements, and must contain at least one <child3> element in that sequence.

Declaring mixed content elements

To declare a mixed content element, you use a content specification that begins with #PCDATA, followed by the pipe character and the child elements allowed for the element. For example:

```
<!ELEMENT para (#PCDATA | term)*>
```

declares that the <para> element can have any number of <term> child elements or parsed character data (PCDATA) as content. You cannot use the occurrence indicators, plus sign (+) and question mark (?), with mixed content elements. The <para> element might appear in an XML document as follows:

```
<para>
    The <term>Semantic Web</term> may become reality.
</para>
```

Declaring elements that can contain any content

It is also possible to declare elements with ANY in the content specification, to specify that the element can contain any content including PCDATA, elements, or a combination of elements and PCDATA. For example:

```
<!ELEMENT name ANY>
```

is an example of such a declaration. This type of declaration is little used and is usually regarded as poor practice.

Attribute declarations

You declare attributes using the <!ATTLIST keyword with the following format:

```
<!ATTLIST element attribute type default>
```

where:

- *element* is the name of the element that owns the attribute
- *attribute* is the name of the attribute: an attribute name can occur only once per element
- *type* defines what the attribute value can contain
- *default* defines whether the attribute is optional and whether it has a default value.

For example:

```
<!ATTLIST name attrib CDATA #REQUIRED>
```

declares that the <name> element has an attribute called attrib, that the attribute value must contain character data (type CDATA), and that the element is required to have the attribute. You can include multiple attributes of an element in a single declaration like this:

```
<!ATTLIST name attrib1 CDATA #REQUIRED
    name attrib2 CDATA #REQUIRED
    name attrib3 CDATA #REQUIRED>
```

Attributes have a different type depending on what kind of values they can have. The different types are described in the following sections.

CDATA attributes

The attribute type CDATA is the type most commonly used for attributes. The values of CDATA attributes can contain character data similar to that allowed for elements with the #PCDATA content specification. (However, CDATA sections, processing instructions and comments are not allowed). Examples of this type of declaration are shown in the previous section.

ID attributes

The attribute type ID specifies that the attribute value must be an *XML name* and must be unique for each element within the document. Attributes of this type are used, for example, as the target for cross-references. The following declaration specifies that the <h2> element must have an id attribute of type ID:

```
<!ATTLIST h2 id ID #REQUIRED>
```

An element can only have one attribute of type ID.

IDREF attributes

Elements with attributes of type IDREF are used to refer to an element with an ID attribute. Attributes of this type are used to provide cross-references, and their value must be an XML name. For example, this declaration specifies that the <link> element must have an xref attribute of type IDREF:

```
<!ATTLIST link xref IDREF #REQUIRED>
```

In an XML document, the id and xref attributes would then be used as follows:

```
<h2 id="a4547"> Interesting Heading</h2>
<para>See <link xref="a4547"/> for more information. </para>
```

Note that because XML names are used for ID and IDREF values, the values of these attributes cannot begin with a numerical character.

IDREFS attributes

The attribute type IDREFS is similar to IDREF, but the attribute value must be a list of XML names separated by whitespace. Attributes of this type are used when an element needs to refer to a number of other elements. Each of the names in the list must be the ID of an element in the document. For example, this declaration specifies that the <link> element must have a team attribute of type IDREFS:

```
<!ATTLIST link team IDREFS #REQUIRED>
```

In an XML document, the `id` and `team` attributes would then be used as follows:

```
<member> id="a2472">Sally Jones</member>
<member> id="a2639">Peter Wrightwell</member>

<para>Members <link xref="a2472 a2639"/>.</para>
```

Enumeration attributes

Another type of attribute, called an *enumeration*, specifies that the value can contain one of a list of possible values, separated by the | character. For example, the following declaration specifies that the attribute value can be one of the values x, y or z:

```
<!ATTLIST element attribute (x | y | z) #REQUIRED>
```

The values in the enumeration must be XML names.

ENTITY attributes

The attribute types ENTITY and ENTITIES are used to provide the name of an *unparsed entity* (ENTITY *type*), or a list of entities separated by spaces (ENTITIES *type*). These unparsed entities are external non-XML files, such as graphics files. For example, the following declaration specifies that the <image> element must have a type attribute whose value is an entity reference:

```
<!ATTLIST image type ENTITY #REQUIRED>
```

When you use attributes of type ENTITY or ENTITIES, the DTD must contain an entity declaration for any entities that are referenced by the attribute, and also a so-called *notation* declaration to define information such as the format of the entity. Notations and their relationship to unparsed entities and ENTITY attributes are explained in *Notation declarations* on page 33. For more information about unparsed entities, see *External unparsed general entities* on page 31.

NOTATION attributes

The type NOTATION identifies the attribute with the names of notations declared in the DTD. Attributes of this type are not used much in practice, but you can use them to associate file types with unparsed entities. This might be necessary if a browser or other application needs to know what kind of files unparsed entities are.

For example, the following declaration specifies that the `type` attribute of the `<image>` element can have one of the image type values, `gif`, `jpeg` or `png`:

```
<!ELEMENT image EMPTY>
<!ATTLIST image type NOTATION (gif | jpeg | png ) #IMPLIED
    src CDATA REQUIRED>
```

There would be corresponding notation declarations for each image type, for example:

```
<!NOTATION gif SYSTEM "image/gif">
<!NOTATION jpg SYSTEM "image/jpg">
<!NOTATION png SYSTEM "image/png">
```

and an entity declaration is required for each image used in a document:

```
<!ENTITY example SYSTEM "xmlexample.jpg" NDATA jpg>
```

You might wonder why you would not use an enumeration attribute to specify the image types: the reason is that the value of an enumeration attribute must be an XML name, which does not allow slash characters, as used in `image/gif` and so on.

In an XML document you might then use the `<image>` element as follows:

```
<image src="xmlexample.jpg" type ="jpg"/>
```

Notations are explained in *Notation declarations* on page 33.

NMTOKEN attributes

The attribute types `NMTOKEN` and `NMTOKENS` specify that the attribute value must contain either a single XML *name token* (`NMTOKEN`), or a list of name tokens separated by spaces (`NMTOKENS`). A name token is similar to an XML name, except that any of the allowed characters can be used as the first character in a name token value. Although this type of attribute is little used, name tokens are useful for declaring attributes that must have numeric values. For example, the following declaration specifies that the attribute value of the `<element>` element must have a value that is a name token:

```
<!ATTLIST element attribute NMTOKEN #REQUIRED>
```

Attribute defaults

The final part of an attribute declaration specifies whether the attribute is required or optional and whether the attribute has a fixed value or a default value:

- The `#REQUIRED` keyword, as used in most of the examples in this chapter, specifies that the element must have the attribute.

  The `#IMPLIED` keyword specifies that the attribute is optional in the XML document. For example, the following declaration specifies that the `title` attribute of the `<name>` element is optional:

  ```
  <!ATTLIST name title CDATA #IMPLIED>
  ```

  If a value for the attribute is not provided in the XML document, an application that processes the XML document can use whatever value it likes for the attribute.

- The `#FIXED` keyword specifies that the attribute has a constant value as supplied in the character string following the `#FIXED` keyword. For example, the following declaration specifies that the `version` attribute of the `<product>` element has a fixed value of 1.0:

  ```
  <!ATTLIST product version CDATA #FIXED "1.0">
  ```

  Processing software assumes the attribute to have the fixed value, even if the value is not supplied for the attribute in the XML document. The `#FIXED` keyword cannot be used with attributes of the ID type.

A character string rather than `#REQUIRED`, `#IMPLIED` or `#FIXED` at the end of the attribute declaration specifies the default value for the attribute. For example, the following declaration specifies that the default value for the `platform` attribute of the `<program>` element is Windows:

```
<!ATTLIST program platform NMTOKEN "Windows">
```

In the XML document, the value for such an attribute is optional: if it is omitted, the value is assumed to be the value supplied in the character string.

Entity declarations

You can declare two main types of entity in a DTD:

- *General entities*, which are declared in a DTD and referenced in XML documents
- *Parameter entities*, which are declared and referenced in a DTD (or set of DTDs) only.

When an XML document is validated, the processing software resolves the entity references in the DTD or XML document, including any external XML files, and includes the entity content in the document.

Entities can be termed *parsed,* which means basically that the entity content is in XML format and can be parsed by processing software, or *unparsed*, which refers to non-XML content such as graphics files. Unparsed entities are not handled directly by processing software, but it is possible to provide information to the processing software about how they should be handled. All parameter entities are parsed.

Entities can also be termed *internal*, if the entity content is contained in the DTD, or *external* if the entity content is contained in a separate file. Hence there are five subtypes of entity, as described in the following sections.

Internal parsed general entities

You use this type of entity for boilerplate text that you want to reuse in different XML documents, or for variable items such as product names, and so on. This ensures that the replacement text appears with the same spelling and punctuation throughout your documents. Maintenance is much easier because you only need to change the entity declaration, rather than every instance of the boilerplate text or variable item in your documents.

You declare internal parsed general entities using the following format:

```
<!ENTITY name "replacement text">
```

For example, you might declare an entity called `prodname` for a product name:

```
<!ENTITY prodname "DocULike">
```

and you can then use `&prodname;` as a place-holder for the replacement text anywhere in the XML document. So, if an XML document contains the following code:

```
<p>The &prodname; User Guide. </p>
```

the output document would contain the following text: ‘The DocULike User Guide’.

External parsed general entities

You use this type of entity for embedding XML files in a master XML document, for example legal notices or chapter files in books.

You declare external parsed general entities using the following format:

```
<!ENTITY name SYSTEM "URL">
```

where *URL* indicates the location of the external XML files. For example, you can declare an entity called `footer` as follows:

```
<!ENTITY footer SYSTEM "/boilerplate/footer.xml">
```

The URL can be a full URL or a relative URL. You specify that the entity is to be included in the XML document simply by adding an entity reference:

```
&footer;
```

External unparsed general entities

You use this type of entity for embedding non-XML content such as pictures and multimedia materials in your documents. This is the only type of unparsed entity, so you can refer to them simply as *unparsed entities.*

You declare unparsed entities using the following format:

```
<!ENTITY name SYSTEM "URL" NDATA file_format>
```

For example, to declare an entity called `workflow` that is associated with a JPEG file called `workflow.jpg`:

```
<!ENTITY workflow SYSTEM "http://www.acme.org/workflow.jpg"
NDATA jpeg>
```

Unparsed entities must be associated with a notation declaration to define the format of the entity. Notations and their relationship to unparsed entities and `ENTITY` attributes are explained in *Notation declarations* on page 33.

Internal parameter entities

You use this type of entity to reuse element and attribute declarations within a DTD. If a number of elements have the same attributes, it is easier and less error-prone to define the attributes once and simply refer to that definition in other declarations.

You define internal parameter entities using this format:

```
<!ENTITY % name "definition">
```

For example, the parameter entity declaration below defines a common set of child elements (`<image>`, `<surname>` and so on) that are reused by means of a parameter entity reference in the declarations for the <x> and <y> elements.

```
<!ENTITY % common_elements "image+, surname, forename, middlename*">

<!ELEMENT x (%common_elements;)>
<!ELEMENT y (%common_elements;, element)>
```

The following parameter entity declaration is for the `%common.attrib` parameter entity in DocBook, which itself uses a number of parameter entity declarations:

```
<!ENTITY % common.attrib
   "%id.attrib;
   %lang.attrib;
   %remap.attrib;
   %xreflabel.attrib;
   %revisionflag.attrib;
   %effectivity.attrib;
   %dir.attrib;
   %xml-base.attrib;
   %local.common.attrib;"
>
```

For example, `%revisionflag;` refers to the parameter entity declaration:

```
<!ENTITY % revisionflag.attrib
   "revisionflag(changed
         |added
         |deleted
         |off)#IMPLIED">
```

The common set of attributes that is defined in the `%common.attrib` parameter entity is used in many of the element declarations in DocBook. For example:

```
<!ATTLIST legalnotice
     %common.attrib;
     %legalnotice.role.attrib;
     %local.legalnotice.attrib;
>
```

External parameter entities

You use this type of parameter entity to include declarations from other DTDs in a main DTD. You can use this mechanism to create a set of modular DTDs and main DTDs that reference some or all of the modular

DTDs as required. Both DocBook and DITA make extensive use of parameter entities in this way.

You define an external parameter entity using the following format:

```
<!ENTITY % name SYSTEM "URL">
```

where *URL* indicates the location of the external file. For example, you can declare an entity called `names`, associated with the `names.dtd` file, as follows:

```
<!ENTITY % names SYSTEM "names.dtd">
```

Elsewhere in the DTD, the parameter is referenced as follows:

```
%names;
```

When processing software accesses the main DTD, the referenced file is included.

Notation declarations

When you use non-XML content such as graphics in your XML documents, you must declare the content as unparsed entities. Unparsed entities cannot be processed by software such as an XML parser, so you must provide additional information about the unparsed entity. This information is called a *notation* and it includes helper information that software can use to process the entity's content. You define a notation typically using the following format:

```
<!NOTATION name SYSTEM "helper information">
```

The name is an identifier that you can refer to in attributes, unparsed entity declarations and processing instructions. The helper information might, for example, identify a file type, or the location of a program that can process the unparsed entity. However, there is actually no standard format for notations, as individual software applications have their own expectations for the format and content of notations.

You use notations if you want to include graphics such as JPEG or PNG files in your documentation. You would declare an unparsed entity and associate that entity with a notation that uses a MIME type such as `image/jpg` to identify the file type of the entity. The following example shows a notation declaration for a JPEG file:

```
<!NOTATION jpeg SYSTEM "image/jpeg">
```

In the corresponding entity declaration, the notation name is used after the NDATA keyword.

```
<!ENTITY workflow SYSTEM "http://www.acme.org/workflow.jpg"
NDATA jpeg>
```

To actually include the graphic file in your document, you must use an element with an ENTITY type attribute whose value is the name of the unparsed entity. Suitable DTD declarations are as follows:

```
<!ELEMENT image EMPTY>

<!ATTLIST image source ENTITY #REQUIRED>
```

and to include the entity in the document:

```
<image source="workflow"/>
```

In this example, whenever a reference is made to the workflow entity within an attribute value, the location of the image file, as specified in the ENTITY declaration, and the notation for the jpeg type, are associated with the entity and passed to the processing application.

As another example, you might want to specify that a particular viewer program is used for a particular type of entity. The following declarations associate a PNG viewer program with PNG images so that an XML software application can use the viewer program to view PNG images.

```
<!NOTATION png SYSTEM "http://www.acme.com/PNGViewer.exe">
<!ENTITY teddy SYSTEM "teddy.png" NDATA png>
```

Assuming suitable element and attribute declarations, you can include the entity in the document as follows:

```
<image source="teddy"/>
```

Comments

You use comments in DTDs to provide documentation for yourself and others. Comments have the same format as they do in XML documents:

```
<!-- comment text -->
```

The comments can contain anything: they are ignored by XML parsers.

Conditional sections in DTDs

You can specify that sections of a DTD are ignored by processing software by creating conditional sections. The keyword INCLUDE specifies that declarations are included, while the keyword IGNORE specifies that declarations are ignored.

In the following example, the element declarations are excluded:

```
<![IGNORE [
<!ELEMENT firstname (#PCDATA)>
<!ELEMENT surname (#PCDATA)>
]]>
```

In the next example, the element declarations are included:

```
<![INCLUDE [
<!ELEMENT firstname (#PCDATA)>
<!ELEMENT surname (#PCDATA)>
]]>
```

Conditional sections must consist of an entire declaration or declarations; fragments are not allowed.

You can control the inclusion of declarations in the DTD simply by changing the IGNORE and INCLUDE keywords on the appropriate sections of the DTD. This is useful during the development and testing of a DTD, and also if you need to customize a DTD.

You can only use conditional sections defined with INCLUDE and IGNORE within the external DTD subset of a document and not in the internal DTD subset. However, you can also use INCLUDE and IGNORE with parameter entities:

```
<!ENTITY % optional_declaration "INCLUDE">
<!ENTITY % optional_declaration "IGNORE">
```

In individual documents, you can then include a parameter entity reference in an internal DTD subset to include declarations as required:

```
<![ %optional_declaration [
<!ELEMENT firstname (#PCDATA)>
<!ELEMENT surname (#PCDATA)>
]]>
```

Associating XML documents with DTDs

As described in Chapter 2, *XML Essentials*, an XML document is associated with a DTD through a document type declaration. The document type declaration can:

- Contain the whole DTD
- Reference a DTD file
- Reference a DTD file (the *external subset*) and contain additional declarations in an *internal subset* of the DTD.

The following example shows a document type declaration that refers to a DTD file and contains an internal subset:

```
<!DOCTYPE chapter SYSTEM "http://www.xmlstore.com/dtds/book.dtd"
[
<!ENTITY booktitle "XML Examples">
]>
```

The root element of the document is specified after the `<!DOCTYPE` keyword, and this is usually followed by the `SYSTEM` keyword and the URL where the DTD is located. The following is another example of a document type declaration:

```
<!DOCTYPE book SYSTEM "file:///usr/local/xml/docbook/4.5/docbook.dtd">
```

You can also use a relative URL, or just specify the file name, if the XML document resides in the same directory as the DTD. For example:

```
<!DOCTYPE book SYSTEM "book.dtd">
```

Sometimes the `PUBLIC` keyword is used instead of the `SYSTEM` keyword in the case of industry-standard DTDs, and a non-specific reference is provided:

```
<!DOCTYPE book PUBLIC "-//OASIS//DTD DocBook XML V4.5//EN"
         "http://www.oasis.open.org/docbook/xml/4.5/docbookx.dtd">
```

However, the processing software might not recognize the reference, so you should also provide a second, specific location, as in the second line of the example, although you do not have to include the `SYSTEM` keyword.

Any declarations in the internal subset for elements or attributes declared in the external subset override the corresponding declarations in the external subset. This is useful if you need to modify the behaviour of your DTD. It is also possible to redefine parameter entities in the internal subset of a DTD to override their definitions in the main DTD.

DTD example

Figure 3.1 contains an example DTD for an ISTC membership list that illustrates many of the types of declaration described in this chapter, and Figure 3.2 on page 38 shows an example XML document that is valid according to the DTD.

Figure 3.1 Example DTD

```
<!--Document Type Definition for ISTC Members-->
<!-- Entity declaration -->
<!ENTITY sig "Special Interest Group">

<!-- Root element -->
<!ELEMENT istc (member+)>

<!-- member element with 5 child elements -->
<!ELEMENT member (membership,name,address,telephone,notes)>
<!-- Required id attribute -->
<!ATTLIST member id ID #REQUIRED>

<!-- membership is an empty element -->
<!ELEMENT membership EMPTY>
<!-- number attribute must have numerical value, grade attribute has three
possible values, with "Member as default, subs_paid value must contain
character data -->

<!ATTLIST membership number NMTOKEN #REQUIRED
            grade (Member|Fellow|Student) "Member"
            subs_paid CDATA #REQUIRED>

<!-- name element with 3 child elements, middle_name is optional -->
<!ELEMENT name (first_name,middle_name?,surname)>
<!-- Optional title attribute -->
<!ATTLIST name title CDATA #IMPLIED>

!-- Element contains parsed character data -->
<!ELEMENT first_name (#PCDATA)> <
<!ELEMENT middle_name (#PCDATA)>
<!ELEMENT surname (#PCDATA)>
```

Figure 3.1 Example DTD (continued)

```
<!-- postcode child element is optional -->
<!ELEMENT address (line1,line2,line3,postcode?)>
<!ELEMENT line1 (#PCDATA)>
<!ELEMENT line2 (#PCDATA)>
<!ELEMENT line3 (#PCDATA)>
<!ELEMENT postcode (#PCDATA)>

<!-- landline and mobile child elements are both optional -->
<!ELEMENT telephone (landline?,mobile?)>
<!ELEMENT landline (#PCDATA)>
<!ELEMENT mobile (#PCDATA)>

<!-- memberref element has ref attribute of type IDREF -->
<!ELEMENT memberref EMPTY>
<!ATTLIST memberref ref IDREF #REQUIRED>

<!-- notes is a mixed content element -->
<!ELEMENT notes (#PCDATA|memberref)*>
```

Figure 3.2 Valid XML document for example DTD

```
<?xml version="1.0"?>
<!DOCTYPE istc SYSTEM "istc.dtd">
<istc>
<member id="m3467">
   <membership number="3467" grade="Fellow" subs_paid=""/>
   <name>
      <first_name>Peter</first_name>
      <middle_name>J</middle_name>
      <surname>Wrightwell</surname>
   </name>
   <address>
      <line1>23, The Willows</line1>
      <line2>Midsommer Worthy</line2>
      <line3>Wessex</line3>
      <postcode>MW1 6HP</postcode>
   </address>
   <telephone>
      <landline>02346 78689</landline>
      <mobile>087 5738945</mobile>
   </telephone>
   <notes>
      On XML &sig; with <memberref ref="m3467"/>.
   </notes>
</member>
</istc>
```

Apart from using a DTD, the other main way of defining an XML language is to use the XML Schema language, also known as XML Schema Definition (XSD), which is itself an XML language developed by the W3C. XML Schema has a number of advantages over DTDs:

- *Strong data typing.* You can specify that the content of an element corresponds to a particular type of data, such as strings, integers, date and time values, or URLs.
- *Tool support.* Because XML Schema uses XML syntax, you can use standard XML authoring and processing tools to work with schemas.
- *Better facilities for constraining XML document content.* For example, you can restrict the number and sequence of elements using XML schema in more ways than is possible with DTDs.
- *Better support for namespaces.* If namespaces are used with a DTD, all elements from the namespace must be declared with their prefix.
- *Reuse.* XML Schema has better facilities for reusing declarations already defined in other schemas.

One disadvantage of XML Schema documents is that they do not handle general entities. Although they can provide the functional equivalent, they do not have the straightforward mechanism for defining and referencing boilerplate text that is provided by DTDs. There are also other advantages to DTDs, including their relatively concise syntax and good tool support.

XML Schema is widely used in software applications, where its ability to constrain data content is invaluable. Despite the advantages that XML Schema offers, DTDs are probably fine for narrative-style documents such as technical publications, although both DocBook and DITA exist in DTD and XML Schema versions.

Like a DTD, an XML Schema document is a means of specifying the elements, attributes and other components that can appear in a particular XML document type.

You define an XML Schema in a schema document, which has the file extension `.xsd`. Within an XML Schema document there are elements that define and constrain the elements that can appear in an XML document based on the schema: such documents are referred to as *instance documents*, as they contain instances of the element and attribute types defined in the schema. The root element of a schema document is `<xs:schema>`, which contains child elements including the following:

- `<xs:element>`: defines the elements that can appear in an instance document.
- `<xs:attribute>`: defines the attributes that can appear in an instance document.
- `<xs:simpleType>`: defines derived simple types, that is, the type of data upon which element content is based.
- `<xs:complexType>`: defines complex types, that is, the type of child elements and attributes that an element can contain.
- `<xs:annotation>`: provides documentation of the XML Schema document.

The example in Figure 3.3 on page 50 shows a simple XML Schema document that is equivalent to the DTD in Figure 3.1 on page 37.

Rather than provide a complete reference, which is beyond the scope of this book, the following sections describe the main features of the XML Schema language. For complete reference information, see for example *XML in a Nutshell.*

The <xs:schema> element

In an XML Schema document you declare elements and attributes using the XML Schema language. Therefore you must specify the namespace of the XML Schema language (`http://www.w3.org/2001/XMLSchema`) and the prefix to be used with the elements and attributes of the XML Schema language, which is usually the `xs:` or `xsd:` prefix. You do this using the `xmlns:xs` attribute of the `<xs:schema>` element.

The elements that you declare in an XML Schema document can belong to a namespace, in which case you use the `targetNamespace` attribute to identify that namespace and an `xmlns` attribute to specify the prefix for the namespace. You can also specify the `elementFormDefault` attribute, to define whether the elements you declare in the XML Schema document must be qualified with a namespace in the instance document. For example:

```
<?xml version="1.0" encoding="utf-8"?>

<xs:schema xmlns:xs="http://www.w3.org/2001/XMLSchema"
targetNamespace="http://www.istc.com/documents/xsd"
xmlns:istc="http://www.istc.com/documents/xsd"
elementFormDefault="qualified">

<!--element declarations skipped -->
</xs:schema>
```

Declaring elements

In an XML Schema document you declare elements using the `<xs:element>` element. Declarations that immediately follow the `<xs:schema>` element are termed top-level elements and can be used as the root element in an XML instance document.

When you declare an element in an XML Schema document, you associate it with a type. The type can be a *simple type*, in which case the element can contain text only and cannot have attributes, or a *complex type*, in which case the element can contain child elements and attributes. The type of an element in XML Schema is equivalent to the content specification of an element declaration in a DTD. The elements in an XML instance document can have only values that fit the types defined for the elements in the schema document.

Simple types

There are a number of built-in simple types, such as `string`, `integer` and `date`. For example, the following markup declares an element `<surname>` that can contain a text string only:

```
<xs:element name="surname" type="xs:string">
</xs:element>
```

The `name` attribute gives the name of the element being declared, while the `type` attribute gives the type, in this case, `xs:string`, one of the predefined types in XML Schema. There are 44 predefined types, covering different types of integer, Boolean values, date and time strings, types corresponding to the attribute types in DTDs, and others. For a description of these types, refer to the appropriate reference chapter in *XML in a Nutshell* or `www.w3.org`.

Apart from the built-in simple types, you can base elements on derived simple types, which are types declared in the XML Schema document and derived by restriction or extension of the allowed content of the simple type. For example, the following declaration is for a derived simple type persType1 that restricts the length of an integer to six digits:

```
<xs:simpleType name="persType1">
   <xs:restriction base="xs:integer">
      <xs:maxLength value="6"/>
   </xs:restriction>
</xs:simpleType>
```

The <xs:restriction> element declares that the integer simple type is to be restricted, and the <xs:maxLength> element determines the maximum length of the derived type.

As another example, the following type declaration is for a string that must consist of uppercase A-Z characters followed by four numeric digits:

```
<xs:simpleType name="persType2">
   <xs:restriction base="xs:string">
      <xs:pattern value="[A-Z][0-9]{4}"/>
   </xs:restriction>
</xs:simpleType>
```

The <xs:restriction> element declares that the string simple type is to be restricted, and the <xs:pattern> element determines the pattern of content that the string can contain.

A derived simple type, as in the second example of the persType2 type, would then be used in an element declaration as follows:

```
<xs:element name="codename" type="xs:persType2">
</xs:element>
```

In a valid instance document, the <codename> element might have the following content:

```
<codename>B1234</codename>
```

However, the following content of the <codename> element would not be valid, as it does not match the specified pattern:

```
<codename>a12346</codename>
```

Another example of restriction is illustrated in the simple type that forms part of the declaration for the <membership> element in the example in Figure 3.3 on page 50:

```
<xs:simpleType>
   <xs:restriction base="xs:token">
      <xs:enumeration value="Member"/>
      <xs:enumeration value="Fellow"/>
      <xs:enumeration value="Student"/>
   </xs:restriction>
</xs:simpleType>
```

This simple type is equivalent to an enumeration type attribute in a DTD.

In the XML Schema language there are many other ways to extend or restrict the content of a simple type.

Complex types

Complex types can contain child elements and attributes. A complex type can be defined as part of an element declaration, or it can be defined as a named type that is reused in many element declarations.

The following example shows a declaration for an <officer> element that is based on a complex type called officerType. The complex type declaration specifies that the <officer> element can contain a sequence of the child elements, <image>, <surname>, <forename> and <middlename>. The complex type declaration also specifies that the <officer> element has the persid attribute.

```
<xs:element name="officer" type ="officerType">
   <xs:complexType name="officerType">
     <xs:sequence>
        <xs:element name="image" maxOccurs="unbounded"/>
        <xs:element name="surname"/>
        <xs:element name="forename"/>
        <xs:element name="middlename" minOccurs="0" maxOccurs="unbounded">
        <xs:complexType>
          <xs:attribute name="persid" type="persType"/>
        </xs:complexType>
        </xs:element>
     </xs:sequence>
   </xs:complexType>
</xs:element>
```

An element declaration such as this can also use the ref attribute with the <xs:element> declarations for child elements. See for example the declaration for the <member> element in Figure 3.3 on page 50.

The declaration for the <middlename> element above shows that you can specify the frequency of elements with the minOccurs and maxOccurs attributes. These attributes have a default value of 1, but can also have the value 0, unbounded, or another integer value.

Table 3.1 shows that values of 0, 1, and unbounded for minOccurs and maxOccurs produce the same effect as the frequency operators in a DTD. However, you can also specify other values of these attributes to define element frequencies that are not possible with DTDs.

Table 3.1 Specifying element frequency in XML Schema documents and DTDs

minOccurs value	*maxOccurs value*	*DTD equivalent*	*Meaning*
0	1	?	Zero or one of the child element is allowed
0	unbounded	*	Zero or more of the child element is allowed
1	unbounded	+	One or more of the child element is allowed
X	Y	No equivalent	The element must have at least *x* child elements and no more than *y* child elements

In a complex type declaration, the <xs:sequence> element specifies the sequence of child elements and provides equivalent functionality to the ',' characters in the content specification of DTDs. The <xs:choice> element specifies that one of a list of child elements must appear, and provides equivalent functionality to the '|' character in the content specification of DTDs.

```
<xs:element name="platform">
  <xs:complexType>
    <xs:choice>
      <xs:element name="Windows"/>
      <xs:element name="UNIX"/>
      <xs:element name="Mac"/>
    </xs:choice>
  </xs:complexType>
</xs:element>
```

There is also an `<xs:all>` element that declares that each child element must appear only once but can appear in any order:

```
<xs:element name="name">
   <xs:complexType>
      <xs:all>
         <xs:element name="given_name"/>
         <xs:element name="middle_name"/>
         <xs:element name="family_name"/>
      </xs:all>
   </xs:complexType>
</xs:element>
```

Defining other content models

The previous sections describe how XML Schema can define two of the types of content specification that are possible with DTDs. This section describes how you can declare mixed content elements and empty elements in XML Schema.

For an empty element:

```
<xs:element name="phone">
   <xs:complexType>
      <xs:attribute name="number" type="xs:string">
   </xs:complexType>
</xs:element>
```

For a mixed content element you can use the `mixed` attribute of the `<complexType>` element. The following example specifies that character data can occur in the content of the element:

```
<xs:element name="skill-level">
   <xs:complexType mixed="true">
      <xs:choice>
         <xs:element name="beginner"/>
         <xs:element name="advanced"/>
      </xs:choice>
   </xs:complexType>
</xs:element>
```

You can use the <xs:any> element to specify that an element can contain any content. For example:

```
<xs:element name="notes" minOccurs="0">
   <xs:complexType>
      <xs:sequence>
         <xs:any namespace="http://www.w3.org/1999/xhtml"
            minOccurs="0" maxOccurs="unbounded"
            processContents="skip"/>
      </xs:sequence>
   </xs:complexType>
</xs:element>
```

In this example, you could include any XHTML elements (as identified by the namespace attribute) within the content of the <notes> element. The processContents attribute specifies whether the markup from the included namespace is to be validated.

Declaring attributes

In an XML Schema document you declare attributes using the <xs:attribute> element. When you declare an attribute in an XML Schema document you associate it with a simple type. You can declare attributes globally as top-level attributes that you can reference from anywhere in the schema, or as part of a complex type declaration, as shown in the previous section.

The following example shows a top-level attribute declaration:

```
<xs:attribute name="persid" type="persType"/>
```

Attribute groups

You can declare groups of attributes using the <xs:attributeGroup> element. This is equivalent to the use of parameter entities in a DTD and has the same advantage of easy maintenance. For example:

```
<xs:attributeGroup name="commonatts">
   <xs:attribute name="firstname" type= "xs:string"/>
   <xs:attribute name="surname" type= "xs:string"/>
</xs:attributeGroup>
```

You can then refer to the group of attributes as in the following example:

```
<xs:element name="personName">
...
  <xs:attributeGroup ref="commonatts"/>
...
</xs:element>
```

Annotations

You can use the <xs:annotation> element in XML Schema documents to provide documentation about the schema. The <xs:annotation> element can be a top-level element or a child of most XML Schema elements. The advantage of using the <xs:annotation> element over adding comments is that the content of <xs:annotation> is more easily machine-readable.

The <xs:annotation> element has as child elements <xs:documentation> and <xs:appinfo>. These child elements are similar, but <xs:documentation> is intended more for human-readable information and <xs:appinfo> is intended for information that can be processed by an application. The <xs:documentation> element can contain any well-formed XML markup, so lends itself to the automatic generation of documentation.

The following is an example of the use of the <xs:documentation> element:

```
<xs:annotation>
  <xs:documentation>
  Schema for error message documentation
  </xs:documentation>
</xs:annotation>
```

Using multiple schema documents

You can split large schema documents into smaller documents so that you can group related declarations into modules. You can then reuse declarations from other schema documents in a particular schema document.

You use the following three elements for this purpose:

- <xs:include> to bring in content from an external schema
- <xs:import> to include declarations from an external schema that belongs to a different namespace
- <xs:redefine> to extend or modify the declarations from an external schema.

In the following DocBook example, the <xs:include> element imports the dbnotnx.xsd document and two other documents and makes all their declarations available for reuse.

```
<xs:schema
  xmlns:xs="http://www.w3.org/2001/XMLSchema"
    elementFormDefault="qualified">

  <xs:include schemaLocation="dbnotnx.xsd"/>
  <xs:include schemaLocation="dbpoolx.xsd"/>
  <xs:include schemaLocation="dbhierx.xsd"/>
...
</xs:schema>
```

In the following example, the top-level <xs:import> element imports the TechPubs.xsd document, and makes all its declarations available for reuse.

```
<xs:import namespace = "http://www.acme.com/documents"
  schemaLocation = "http://file_Location/TechPubs.xsd"/>
```

The namespace attribute value contains the namespace URI for the imported schema, and the schemaLocation attribute contains a relative or absolute URL pointing to the actual location of the schema document to import.

It can be useful to redefine declarations brought in using the <xs:import> or <xs:include> elements, for example to extend or restrict a type definition. The following illustrates an <xs:redefine> element that redefines a declaration from the dbnotnx.xsd schema document:

```
<xs:redefine schemaLocation="dbnotnx.xsd">
<!--declaration -->
</xs:redefine>
```

These facilities for reuse in XML Schema are comparable to the use of parameter entities in DTDs.

Notations in XML Schema

In XML, notations are used to declare links to external non-XML content such as graphics files and to associate the content with software applications that handle it. In XML Schema you declare notations using the `<xs:notation>` element, which is the equivalent of a `<!NOTATION>` declaration in a DTD. Notations must also be associated with a simple type derived from the built-in `xs:NOTATION` type, as shown in the following example:

```
<xs:notation name="jpeg" public= "image/jpeg" system = "JPEG_Viewer.exe" />
<xs:notation name="gif" public= "image/gif" system = "GIF_Viewer.exe" />

<xs:simpleType name="notation.Image">
  <xs:restriction base="xs:NOTATION">
    <xs:enumeration value="jpeg"/>
    <xs:enumeration value="gif"/>
  </xs:restriction>
</xs:simpleType>
```

Associating XML documents with XML Schema documents

If the XML Schema document does not have a namespace, you can use markup in the instance document to specify the location of the XML Schema document, as shown in this example:

```
<?xml version="1.0" encoding="utf-8"?>
<messages xmlns:xsi="http://www.w3.org/2001/XMLSchema-instance"
  xsi:noNamespaceSchemaLocation="C:\MessageXML\Errormess.xsd">
...
</messages>
```

The `xmlns:xsi` attribute associates the instance document with the official XML Schema instance namespace `http://www.w3.org/2001/XMLSchema-instance`, and in this case the attribute:

```
xsi:noNamespaceSchemaLocation
```

specifies the location of the XML Schema document.

If the XML Schema document does have a namespace, you must use the `xmlns` and `xsi:schemaLocation` attributes to specify the namespace to use, and the location of the XML Schema document. For example:

```
<?xml version="1.0" encoding="utf-8"?>

<messages xmlns="http://www.cowan.com"
  xmlns:xsi="http://www.w3.org/2001/XMLSchema-instance"
  xsi:schemaLocation="http://www.cowan.com Errormess.xsd">
  ...
</messages>
```

Example XML Schema document

The XML Schema document shown in Figure 3.3 is equivalent to the DTD in Figure 3.1 on page 37.

Figure 3.3 Example XML Schema document

```
<?xml version="1.0" encoding="UTF-8"?>
<!-- XML Schema Document for ISTC Members -->
<xs:schema xmlns:xs="http://www.w3.org/2001/XMLSchema"
  elementFormDefault="qualified">

  <xs:element name="istc">
    <xs:complexType>
      <xs:sequence>
        <xs:element maxOccurs="unbounded" ref="member"/>
      </xs:sequence>
    </xs:complexType>
  </xs:element>

  <xs:element name="member">
    <xs:complexType>
      <xs:sequence>
        <xs:element ref="membership"/>
        <xs:element ref="name"/>
        <xs:element ref="address"/>
        <xs:element ref="telephone"/>
        <xs:element ref="notes"/>
      </xs:sequence>
      <xs:attribute name="id" use="required" type="xs:ID"/>
    </xs:complexType>
  </xs:element>
```

Figure 3.3 Example XML Schema document (continued)

```
<xs:element name="membership">
  <xs:complexType>
    <xs:attribute name="number" use="required" type="xs:NMTOKEN"/>
    <xs:attribute name="grade" default="Member">
      <xs:simpleType>
        <xs:restriction base="xs:token">
          <xs:enumeration value="Member"/>
          <xs:enumeration value="Fellow"/>
          <xs:enumeration value="Student"/>
        </xs:restriction>
      </xs:simpleType>
    </xs:attribute>
    <xs:attribute name="subs_paid" use="required"/>
  </xs:complexType>
</xs:element>

<xs:element name="name">
  <xs:complexType>
    <xs:sequence>
      <xs:element ref="first_name"/>
      <xs:element minOccurs="0" ref="middle_name"/>
      <xs:element ref="surname"/>
    </xs:sequence>
    <xs:attribute name="title"/>
  </xs:complexType>
</xs:element>

<xs:element name="first_name" type="xs:string"/>
<xs:element name="middle_name" type="xs:string"/>
<xs:element name="surname" type="xs:string"/>

<xs:element name="address">
  <xs:complexType>
    <xs:sequence>
      <xs:element ref="line1"/>
      <xs:element ref="line2"/>
      <xs:element ref="line3"/>
      <xs:element minOccurs="0" ref="postcode"/>
    </xs:sequence>
  </xs:complexType>
</xs:element>

<xs:element name="line1" type="xs:string"/>
<xs:element name="line2" type="xs:string"/>
<xs:element name="line3" type="xs:string"/>
<xs:element name="postcode" type="xs:string"/>
```

Figure 3.3 Example XML Schema document (continued)

```
    <xs:element name="telephone">
      <xs:complexType>
        <xs:sequence>
          <xs:element minOccurs="0" ref="landline"/>
          <xs:element minOccurs="0" ref="mobile"/>
        </xs:sequence>
      </xs:complexType>
    </xs:element>
    <xs:element name="landline" type="xs:string"/>
    <xs:element name="mobile" type="xs:string"/>

    <xs:element name="memberref">
      <xs:complexType>
        <xs:attribute name="ref" use="required" type="xs:IDREF"/>
      </xs:complexType>
    </xs:element>

    <xs:element name="notes">
      <xs:complexType mixed="true">
        <xs:sequence>
          <xs:element minOccurs="0" maxOccurs="unbounded" ref="memberref"/>
        </xs:sequence>
      </xs:complexType>
    </xs:element>
  </xs:schema>
```

Parsing and validating XML documents

To analyse and process an XML document, you need software called an *XML parser*. Web browsers incorporate XML parsers, as do software programs that transform XML into other formats. XML editors also include XML parsers, so that they can validate and display source XML documents with appropriate formatting. Unless you are involved in the programming aspects of XML processing, you need not be aware of specific XML parsers, such as Xerces from the Apache XML project, or Microsoft's MSMXL, which is built into Internet Explorer. However, it is useful to know a little about how this type of software is used.

XML parsers do the following:

- Check well-formedness
- Check validity (if they are validating parsers)
- Resolve entities declared in DTDs
- Associate unparsed entities with URLs and notations

- Supply default attribute values
- Perform other processing.

After this processing, the parser passes on a document or data to the software program, such as a browser or XML editor, that processes or uses the data further.

It is important to validate XML documents to ensure that they have correct structure. Most if not all XML parsers are *validating parsers*, that is, they check that XML documents conform to their associated DTD or schema. All available XML parsers support validation against DTDs: however, not all of them support validation against XML Schema documents. Parsers such as MSMXL also often include an XSLT engine for transforming XML documents: however, parsers vary in their support for XSLT, not only in terms of whether they support it at all, but also in terms of which parts of the specification they support. Support for namespaces also varies between parsers.

XML parsers can take different approaches to processing an XML document:

- *Event-driven approach.* Each statement of the XML document is processed in turn. Simple API for XML (SAX) is a standard application programming interface (API) used by parsers for this approach.
- *Tree-based approach.* In this type of parser a tree structure representing the structure of the XML document is built in memory. The Document Object Model (DOM) API is an API used by parsers for this approach.

The method used depends on how the XML parser needs to process the XML document. An event-driven approach uses less memory than a tree-based approach, is often faster, and can be the only approach for very large XML documents that cannot fit into memory. However, the event-driven approach cannot be used for XML validation that requires access to particular parts of the XML document at a given time. For example, some types of XML processing, such as XSLT processing, require access to the entire XML document in memory. In fact, most parsers (including Xerces and MSMXL) can use both approaches, as they support both the SAX and DOM APIs.

4

XML Documentation Languages

This chapter discusses XML languages and standards that are designed specifically for producing technical documentation: DocBook, Darwin Information Typed Architecture (DITA) and S1000D.

DocBook, DITA, and S1000D allow you to create documents with structured content that can be published to a variety of formats, with the appropriate presentation information being applied at publishing time:

For each of the languages, this chapter describes:

- The features of the language and its advantages for technical documentation
- The markup of the language
- The tools available for working with the language.

There are also sections comparing DocBook with DITA, and S1000D with DITA.

DocBook

DocBook is a markup language for describing books, papers, articles and other forms of documentation that is well suited to hardware and software documentation. DocBook began as an SGML application designed by HaL Computer Systems and O'Reilly & Associates around 1991, and later a DocBook XML application was developed. The first official XML version of DocBook was Version 4.1.2, and DocBook Version 4.5 became a standard of the Organization for the Advancement of Structured Information Standards (OASIS). OASIS is a not-for-profit consortium that fosters the development and adoption of open standards such as DocBook and more recently DITA. The DocBook Technical Committee of

OASIS is responsible for maintaining the DTDs and XML schemas for DocBook.

DocBook resources, including schemas and documentation, are available from `www.docbook.org`.

DTD, XLM Schema and RELAX NG implementations of DocBook are available for the different versions of DocBook. The examples and descriptions in this chapter are applicable to the DTD implementation of DocBook Version 4.5.

DocBook markup

The DocBook document illustrated in the following example shows how part of the content of a book could be marked up in the DocBook language.

```
<?xml version="1.0" encoding="utf-8"?>
<!DOCTYPE book PUBLIC "-//OASIS//DTD DocBook XML V4.5//EN"
"http://www.oasis-open.org/docbook/xml/4.5/docbookx.dtd">
<book id="simple_book">
<bookinfo>
  <title>My Life With XML</title>
  <author><firstname>Peter</firstname><surname>Wrightwell</surname></author>
  <copyright><year>2008</year><holder>Peter Wrightwell</holder></copyright>
</bookinfo>
<preface><title>Prologue</title> ... </preface>
<chapter id="chapter_1"><title>How I Discovered XML</title>
  <para>This chapter is about my first encounter with XML.</para>
  <sect1><title>Early Markup</title>
    <para> From my early years I longed for an extensible markup
    language.</para>
    <para> … </para>
    <example> ... </example>
  </sect1>
</chapter>
<chapter id="chapter_2"> ... </chapter>
<appendix> ... </appendix>
<index> ... </index>
</book>
```

The example shows that DocBook has elements that reflect the organization of a book: however, DocBook also has elements for other types of documentation. Apart from books, DocBook lends itself to authoring articles and reference pages. The reference page or manual page in DocBook is designed to mimic the UNIX ‘manpage’ concept.

You could author a complete book in one source file, but with DocBook you typically break a document into separate chunks by using entity declarations. For example, for a book with three chapters and an appendix, you can create a master file such as the following:

```
<?xml version="1.0" encoding="utf-8"?>
<!DOCTYPE book PUBLIC "-//OASIS//DTD DocBook XML V4.5//EN"
"http://www.oasis-open.org/docbook/xml/4.5/docbookx.dtd"
[
<!ENTITY chapter1 SYSTEM "chap1.xml">
<!ENTITY chapter2 SYSTEM "chap2.xml">
<!ENTITY chapter3 SYSTEM "chap3.xml">
<!ENTITY appendixa SYSTEM "appa.xml">
]>
<book><title>My Life With XML</title>
&chapter1;
&chapter2;
&chapter3;
&appendixa;
</book>
```

This usage of entities allows you to have the chapters and appendices in separate files. For example, Chapter 1 might begin like this:

```
<chapter id="ch1"><title>My First Chapter</title>
<para>My first paragraph.</para>
...
</chapter>
```

Note that the chapter files do not have document type declarations.

The elements of DocBook

DocBook is an extensive language with over 400 elements. Table 4.1 summarizes the different categories of element that are available.

There are many more elements than those listed in the table, including a number of elements for producing indexes - so many in fact that new users can find DocBook overwhelming. *DocBook: The Definitive Guide* includes a complete reference on DocBook elements and attributes - see Appendix B, *Bibliography*. There is also a smaller subset of DocBook Version 4.5 called *Simplified DocBook* that has about 100 elements.

Table 4.1 DocBook element categories

Category of element	*Example of elements*	*Description*
Sets	`<set>`	Used for a series of books on a single subject that you want to access and maintain as a single unit.
Book	`<book>`	The top-level element of a book, which can contain divisions and parts, as well as `<toc>` and `<index>` elements.
Divisions	`<division>`	Optional subdivision of books that contains part elements.
Parts	`<part>`	Optional subdivision of books that contains component elements.

Components	`<chapter>`
	`<appendix>`
	`<glossary>`
	`<bibliography>`
		Subdivisions of books and parts that contain chapter-like content.

Sections	`<section>`
	`<glossdiv>`
	`<bibliodiv>`
		Subdivisions of components.

Meta-information elements	`<author>`
	`<title>`
	`<publisher>`
		Child elements of section-level elements and above that provide information about a document, as opposed to the content of a document itself.

Block elements	Elements occurring below component and section-level elements that are roughly equivalent to paragraph level elements, and which define blocks of content.	

	`<itemizedlist>`
	`<orderedlist>`
	`<simplelist>`
		Elements for lists.

	`<caution>`
	`<important>`
	`<note>`
	`<tip>`
	`<warning>`
		Elements for admonitions and notes.

	`<address>`
	`<programlisting>`
	`<screen>`
		Elements that preserve white space and line breaks as in, for example, addresses and program listings.

Category of element	*Example of elements*	*Description*
	`<example>` `<figure>` `<table>`	Elements for examples, figures, and tables.
	`<para>` `<simpara>`	Elements for different types of paragraph.
	`<equation>` `<informalequation>`	Elements for mathematical equations.
	`<alt>` `<graphic>` `<mediaobject>`	Elements for graphics. The alternative text element `<alt>` may also be an inline element.
	`<qandaset>`	Elements for questions and answers collections, such as FAQs and questionnaires.
	`<blockquote>` `<msgset>` `<procedure>`	Miscellaneous block elements, such as those for sets of error messages and procedures.
Inline elements		Elements for marking up running text.
	`<emphasis>` `<phrase>` `<quote>` `<trademark>`	Elements representing traditional publishing text items.
	`<anchor>` `<link>` `<xref>`	Elements used for cross-references.
	`<computeroutput>` `<userinput>`	Elements used for special presentation such as computer output and user input.
	`<inlineequation>` `<subscript>` `<superscript>`	Elements for mathematical equations.
	`<guibutton>` `<guiicon>` `<guimenu>`	Elements for use interface items.
	`<command>` `<errorcode>` `<function>` `<parameter>`	Elements for programming languages and constructs and other items related to programming such as error messages.

Table 4.1 DocBook element categories (continued)

Category of element	Example of elements	Description
	<application> <systemitem>	Elements pertaining to operating systems or operating environments.
	<filename> <hardware> <symbol>	General purpose elements.

DocBook common attributes

DocBook has a number of common attributes that can be used on almost all elements. They are mainly used in conditional processing, in cross-referencing, and for revision marking:

- `arch`: the computer or chip architecture to which the element applies.
- `conformance`: the standards conformance characteristics of the item contained in the element, for example `lsb` (Linux Standards Base).
- `id`: a unique (at least within a document) identifying string for the element.
- `lang`: a language code such as `de_DE` for German, used, for example, to signal to processing applications to change hyphenation and other display characteristics.
- `os`: the operating system to which the element is applicable.
- `remap`: an element name or similar semantic identifier assigned to the content in a previous markup scheme.
- `revision`: the editorial revision to which the element belongs.
- `revisionflag`: the revision status of the element; which can be one of the values: changed, added, deleted, or off.
- `role`: a string used to classify an element.
- `userlevel`: the level of user experience to which the element applies, for example, beginner or advanced.
- `vendor`: the product vendor to which the element applies.
- `xreflabel`: the text to be used when a cross reference (xref) is made to the element.

Later versions of DocBook have added additional common attributes.

The DocBook DTD

The DocBook DTD, available from OASIS, consists of a collection of declarations stored in a number of modules. For example:

- `dbhier.mod`. The module that declares the elements that provide the hierarchical structure of DocBook (sets, books, chapters and so on).
- `dbpool.mod`. The information pool module that declares the elements that describe content (inline elements, bibliographic data and so on) but which are not part of the large-scale hierarchy of a document.
- `dbnotn.mod`. The module that declares the notations used by DocBook.
- `dbcent.mod`. The module that declares and references the ISO character entity sets used by DocBook.
- `dbgenent.mod`. The module that declares general entities for use in DocBook instances. This is the place to add, for example, boilerplate text and organization logos.
- `Cals-tbl.dtd`. A module with declarations for the CALS Table Model, an initiative by the United States Department of Defense to standardize the document types used across branches of the military. The CALS table model, published in MIL-HDBK-28001, was for a long time the most widely supported SGML table model.

There are also:

- Entity (`.ent`) files containing XML entity declarations for characters. The character entity sets distributed with DocBook XML are direct copies of the official entities located at `www.w3.org/2003/entities`.
- A catalog file, as described in *XML catalog files and DocBook* on page 65.

The main DTD file is `docbook.dtd`, which declares and references the other modules. In `docbook.dtd` the modules are combined by using parameter entities that reference the modules:

```
<!ENTITY % dbpool SYSTEM "dbpool.mod">
<!ENTITY % dbhier SYSTEM "dbhier.mod">
%dbpool;
%dbhier;
```

So, for example, the parameter entity `dbpool` is associated with the file `dbpool.mod`.

Customizing DocBook

One powerful feature of DocBook is its ability for customization, the main reason being to add industry-specific elements. For example, you might want to add elements for describing constructs of a particular programming language.

You can customize DocBook in the following ways:

- Add and remove elements
- Change the structure of existing elements
- Add and remove attributes
- Broaden the range of values allowed in an attribute
- Narrow the range of values in an attribute to a specific list or a fixed value.

To customize DocBook, you can write a DTD that references some or all of the DocBook modules, and modify the DTD by adding your own declarations that override declarations in the DTD modules. This is called a *customization layer*. Modifying the DTD in this way avoids having to edit DocBook modules directly, which makes maintaining the modules easier. If newer versions of DocBook are published, you then only have to make changes to your customization layer to keep the modification in step with the new version.

Some people argue against customizing DocBook, because if you extend DocBook by adding new structures, you lose the benefit of an interchangeable format that allows information to be formatted in a vendor-independent fashion. This can create a problem for other people with whom you share documents. Also, extending DocBook may create problems with the tools and stylesheets that you use: strictly speaking, the extended language is no longer DocBook. Creating a DTD that is a strict subset of DocBook is less of a problem, as you can share documents and your XML documents are valid DocBook instances.

If you do modify the structure of the DTD, you must change the public identifier that you use for the DTD and the modules you changed. The license agreement under which DocBook is distributed gives you complete freedom to change, modify and reuse the DTD in any way you like, but you must not call your alterations 'DocBook'.

Conditional processing with DocBook

In technical communication, you often need to produce different versions of a document for different audiences, different platforms, different vendors, and so on. For example, you might need different versions of a document because different installation instructions apply for Windows and for UNIX.

DocBook allows you to identify content in a document that is applicable for particular audiences, platforms and other criteria. When the document is processed, through the use of a stylesheet, such conditional content can be included or excluded as required. Conditional processing is useful for producing multiple versions of a document from a single source document, with the advantage of reduced maintenance and production costs.

You can use a number of DocBook's common attributes to identify conditional content: for example, the `arch`, `conformance`, `os`, `userlevel` and `vendor` attributes. To identify an element (and its children) as having conditional content, you add one of more attributes with appropriate values to the relevant element. For example, you could use the `os` attribute to identify sections as applicable to either Windows or UNIX, as follows:

```
<sect1 os="win">
  <title>Installing on Windows</title>
  ...
</sect1>
<sect1 os="unix">
  <title>Installing on UNIX</title>
  ...
</sect1>
```

You would then process the document with different stylesheets depending on whether you wanted to produce Windows or UNIX versions of the document. You must select the values that are used on the attribute (such as `win` and `unix` in the example), and of course you must be consistent in their usage.

For more information about processing of DocBook documents, see *Publishing from DocBook* on page 64.

Tools for working with DocBook

You can use both simple text editors and XML editors with DocBook files. XML editors can automatically validate your files: however, if you use a simple editor, you must use a separate validating parser. A number of free parsers are available, such as the Xerces-J (formerly XML4J) parser available from xerces.apache.org/xerces2-j/ and the XP tool available from www.jclark.com.

With over 400 elements DocBook is not a simple language, which makes it difficult to have drop-downs for selecting elements in XML editing tools. On the other hand, *validating editors* only present elements that are valid at a particular point in a document. A list of such editors that support DocBook is given in wiki.docbook.org/topic/DocBookAuthoringTools.

A structured application for DocBook XML is shipped with FrameMaker Version 7.0 and later (different versions of DocBook are supported for different versions of FrameMaker). A structured application is a set of files that specify how a particular application translates markup documents to and from FrameMaker. The structured application allows you to import and export DocBook XML files in FrameMaker, a process that is known as *round-tripping*.

More information about authoring XML is provided in Chapter 5, *Authoring with XML*.

Publishing from DocBook

You can publish DocBook source documents to a variety of output formats using various readily available tools. Document Style Semantics and Specification Language (DSSSL) is a stylesheet language that has traditionally been used to transform both XML and SGML DocBook to print and output formats, and more recently XSL stylesheets and Cascading Style Sheets (CSS) have been used to transform DocBook XML.

Jade is a free tool that applies DSSSL stylesheets to XML documents to produce RTF, TeX, MIF and SGML formats. For more information about DSSSL and the Jade tool, see www.jclark.com.

The DocBook Project development team (docbook.sourceforge.net) maintain a set of XSL stylesheets and a set of legacy DSSSL stylesheets that generate HTML, FO/PDF and other formats including RTF, man pages and HTML Help.

There are many other tools available for editing, processing and publishing with DocBook. Refer to docbook.sourceforge.net for listings of such tools.

The advantages of DocBook

The advantages of DocBook include its open, non-proprietary nature: it can be obtained free of charge and DocBook files are easy to port across environments.

You can edit DocBook source files using any text editor, and support is also provided in more sophisticated authoring tools. Free tools are available for converting from other formats to DocBook. Like other XML languages, DocBook is suitable for single-source publishing, being easily transformable to a variety of output formats, and free tools are also available for doing so.

DocBook is well tried and tested, having been implemented extensively around the world. There is a network of thousand of users and help is available from mailing lists and various websites.

XML catalog files and DocBook

As mentioned in Chapter 3, *Defining XML Languages*, you can use `PUBLIC` identifiers in the `DOCTYPE` declarations in your XML documents, and a `SYSTEM` identifier to find the DTD file (or system entity files) on the local machine. However, if the location of the DTDs changes, you have to update all the XML documents.

To avoid this problem, you can use an XML catalog file. An XML catalog file maps the `PUBLIC` identifier to a specific location on a given machine. This allows you to use stable `PUBLIC` identifiers in the `DOCTYPE` declarations in your documents and not worry about using `SYSTEM` identifiers. If you move the DTD to a new machine, you then only need to edit the catalog file's mapping of the `PUBLIC` identifier, rather than edit all your XML documents.

The following shows part of the catalog file available for DocBook Version 4.5:

```
<public publicId="-//OASIS//DTD DocBook XML V4.5//EN"
  uri="docbookx.dtd"/>

<system systemId="http://www.oasis-open.org/docbook/xml/4.5/docbookx.dtd"
  uri="docbookx.dtd"/>

<system systemId="http://docbook.org/xml/4.5/docbookx.dtd"
  uri="docbookx.dtd"/>
```

You need to edit the file so that the `uri` attribute of the `<system>` element points to the location of your DTD file.

Catalog files are used with other XML languages, including DITA. You can use a catalog to locate not only the DTD, but also entity files and stylesheets used during processing.

DITA – Darwin Information Typing Architecture

DITA is an XML framework for the production of *topic-oriented* technical documentation. Examining the components of DITA's name helps in understanding its purpose:

- *Darwin.* DITA topics correspond to information types that inherit characteristics from other information types and which can be specialized, hence the analogy to Charles Darwin's theories of evolution.
- *Information typing.* In DITA, you develop topics based on types of content, such as concept, task and reference information.
- *Architecture.* DITA is intended to be more than a language, but also an architecture for designing information that encapsulates best practices and extensible design.

DITA was first developed by IBM in the late 1990s, but moved into the public domain when it was donated to OASIS: DITA Version 1.0 became one of their official standards in 2005, and Version 1.1 became a standard in 2007. OASIS is responsible for maintaining the DITA specification and associated tools, such as the DITA Open Toolkit. The DITA specification includes both DTD and XML Schema representations of the DITA architecture, and a language reference that provides explanations of each element in the base DITA information types. The DITA Open Toolkit is a collection of sample files, stylesheets and tools for working with DITA.

DITA provides a topic-oriented means of delivering information that corresponds to the modern documentation philosophy of information typing – see for example `docs.oasis-open.org/dita/v1.0/archspec/infotypes.html`. In this documentation model, you produce different types of information according to the way people use products, so in DITA you develop information in topics that are structured according to their information type. There are topic types for conceptual, task and reference information, and it is possible to build on the basic information types to create new, specialized information types. The topic-oriented nature of DITA provides great flexibility in publishing and reusing content in different deliverables.

DITA topics

In DITA you develop content in small self-contained files, *topics*, that have the `.xml` or `.dita` file extension. One definition of a topic is:

> *A unit of information with a title and content, short enough to be specific to a single subject or answer a single question, but long enough to make sense on its own and be authored as a unit.*

(`dita-ot.sourceforge.net/doc/ot-userguide13/xhtml/faqs/topic_authoring.html`)

There are three core topic types based on a generic topic type (in other words, these topic types are *specializations* of the generic topic type):

- *Concept.* Contains information for explaining a concept, defining a term or describing why a user should perform a task.
- *Task.* Contains step-by-step procedural information for performing a task.
- *Reference.* Contains information about items such as commands, programming keywords or error messages.

Apart from the basic topic types, you can define your own. This process of specialization creates new information types. DITA has an object-oriented mechanism in which specialized topics inherit properties from the generic topic type. For more information, see *DITA specialization* on page 77.

The generic topic type

The generic topic type is the root type for all topics in DITA, the type from which all other types are derived. The generic topic type is used for topics that do not have a specific information type.

A topic contains a `<topic>` element that has a required `id` attribute and contains all other elements. All topics have the same basic structure, regardless of the topic type:

- *Title.* The `<title>` element containing the subject of the topic. The optional `<titlealts>` element can provide different text for navigation or search.
- *Short description.* An optional `<shortdesc>` element, used both in the topic content and in generated summaries that include the topic.

- *Prolog.* An optional `<prolog>` container element for various kinds of topic metadata, such as change history, audience, product and so on.
- *Body.* The `<body>` element containing the actual topic content. While all topic types have the same elements for title, short description and prolog, they each allow different content in their body specific to their information type:
  - *Sections and examples.* Sections and examples that can contain block-level elements such as paragraphs, phrase-level elements or text.
  - *Block-level elements.* Paragraphs, lists and tables that might contain other blocks, phrases or text.
  - *Phrases and keywords.* Phrases that can contain other phrases and keywords as well as text, and keywords that can only contain text.
  - *Images and multimedia.* The `<image>` element is used to include graphics and the `<object>` element is used to include multimedia objects. The <alt> child element of <image> is used to provide text descriptions of graphics for the visually impaired.
- *Related links.* Links to other topics. Rather than embedding links directly in each related topic, it is better practice to use *DITA maps* (see page 72) to define and manage links between topics.
- *Nested topics.* Topics defined inside other topics. You should use nesting carefully, as it can result in complex documents that are less usable.

The example given in *The Task topic type* on page 70 shows a task topic that includes a title, short description, prolog and body section.

DITA domains The DITA specification defines *domains*, which are sets of elements associated with a particular subject area or authoring requirement. The elements in a domain are defined in a domain module. The domain module can be integrated with a topic type to make its elements available to a topic of that type.

The following domains are provided in DITA:

- *Typography.* Elements for highlighting, such as bold and italic.
- *Programming.* Elements describing programming syntax and programming structures.
- *Software.* Elements describing software program operation.
- *User Interface.* Elements for describing user interface items.

- *Utility.* Elements for document image maps and other useful structures.

DITA attributes

Most attributes of DITA elements belong to the following categories:

- *Identity attributes.* Attributes that identify content for retrieval or linking. This includes the `id` attribute on topics, which must have a unique value, the `id` attribute on elements, and the `conref` attribute that provides a mechanism for reusing fragments of content. See *Reusing elements within topics* on page 80.
- *Metadata attributes.* Attributes that provide additional information about the content. This includes attributes that are used to show change history and to filter content for conditional processing. See *Filtering content for reuse* on page 81.
- *Architectural attributes.* Attributes that provide a mechanism for specialization. This includes:
  - `class`. The attribute that identifies the specialization module for the element type, as well as their ancestor element types and the specialization modules.
  - `domains`. The attribute that lists the names of the domains in use by the current document type and the ancestry for each domain.

For more information about these attributes, see *DITA specialization* on page 77.

The Concept topic type

A *Concept* topic contains conceptual information rather than procedural information, including information that supports user tasks, definitions of terms, rules and guidelines.

An example of markup for a concept topic is as follows:

```
<?xml version="1.0" encoding="utf-8"?>
<!DOCTYPE concept PUBLIC "-//OASIS//DTD DITA Concept//EN"
"../dtd/concept.dtd">
<concept id="ditaattr" xml:lang="en-us">
   <title>DITA Attributes</title>
   <conbody>
      <p>The main types of attribute in DITA are:
         <ul>
            <li>Identity attributes
            <li>Metadata attributes
            <li>Architectural attributes
         </ul>
      </p>
   </conbody>
</concept>
```

The <conbody> element is the main body-level element for a concept. It can contain paragraphs, lists and other elements, as well as sections and examples.

The Task topic type

A *Task* topic contains a procedure describing how to perform a task, together with any supporting contextual and prerequisite information.

An example of markup for a task topic is as follows:

```
<?xml version="1.0" encoding="utf-8"?>
<!DOCTYPE task PUBLIC "-//OASIS//DTD DITA Task//EN"
"../dtd/task.dtd">
<task id="resetpw" xml:lang="en-us">
   <title>Resetting a User's Password</title>
      <titlealts>
         <navtitle>Resetting a Password</navtitle>
      </titlealts>
   <shortdesc>Use this task to reset a password for a user.</shortdesc>
   <prolog>
      <metadata>
         <audience type="admin" experiencelevel="intermediate"/>
      </metadata>
   </prolog>
   <taskbody>
      <prereq>To reset a user's password you must have administrator
         privileges.</prereq>
      <context>You reset a user's password when it has expired or when you
         receive a request from the user to change the password.
      </context>
      <steps>
         <step><cmd>Go to the Change Password page.</cmd></step>
         <step><cmd>Search for the required user ID.</cmd></step>
         <step><cmd>Enter the new password for the user .</cmd></step>
         <step><cmd>Click OK. </cmd></step>
      </steps>
      <result>The user is sent an email containing the new password.</result>
   </taskbody>
</task>
```

The <taskbody> element is the main body-level element inside a task topic. A task body has a specific structure, with elements in this order:

- <prereq>: information needed before starting the current task.
- <context>: background information to help the user understand the task's purpose and outcome.
- <steps>: the main content of the task topic, containing one or more <step> elements for each step in the task.

- <step>: an action that a user must follow to accomplish a task. Each step must contain a command <cmd> element that describes the particular action the user must perform to accomplish the overall task. The step element can also contain <info>, <substeps>, <tutorialinfo>, <stepxmp>, <choices> or <stepresult> child elements, although these are optional.
- <result>: describes the expected outcome for the overall task.
- <example> (optional): provides an example that illustrates or supports the task.
- <postreq> (optional): describes steps or tasks that the user should do after the successful completion of the current task.

The Reference topic type

A *Reference* topic is intended for topics that describe command syntax, programming instructions and other reference material such as error messages: they usually contain detailed, factual material.

An example of markup for a reference topic is as follows:

```
<?xml version="1.0" encoding="utf-8"?>
<!DOCTYPE reference PUBLIC "-//OASIS//DTD DITA Reference//EN"
"../dtd/reference.dtd">
<reference id="refexample" xml:lang="en-us">
  <title>A reference topic/title>
  <refbody>
    <refsyn>Description of command or API syntax
    </refsyn>
    <section> <title>Section title</title> </section>
    <properties>
      <property>
        <proptype>type</proptype>
        <propvalue>value</propvalue>
        <propdesc>description</propdesc>
      </property>
    </properties>
  </refbody>
</reference>
```

The <refbody> element is the main body-level element inside a reference topic. Within the reference body, all child elements are optional and the structure is limited to tables, property lists, syntax sections and generic sections and examples.

DITA DTDs and schemas

The DITA DTDs available from OASIS include modules for each information type and for each domain:

- DTD structural module files: `*typename*.mod`, for example, `task.mod`.
- DTD domain module files: *domainname*`Domain.mod` and *domainname*`Domain.ent`, for example `ProgrammingDomain.mod` and `ProgrammingDomain.ent`.

In addition there are map and attribute DTDs, as well as catalog files. The `ditabase.dtd` file contains parameter entities for all of the information types and domains that you use.

Similarly, the DITA XML Schema consists of separate schema files for information types and for domains:

- Schema structural module files: *typename*`Mod.xsd` and *typename*`Grp.xsd`
- Schema domain module files: *domainname*`Domain.xsd`.

DITA maps

You develop information in DITA as a set of topics, but how do you organize those topics into a meaningful collection for publication? You do so with *DITA maps.*

A DITA map contains a set of references to DITA topics that allows you to organize the topics into a hierarchy and define the relationships between them. The map does not embed topics. For example, a map or a set of maps can define:

- The hierarchy of topics for a table of contents
- The relationship between topics for online navigation
- The collection of topics to be used to publish a deliverable such as a PDF file.

DITA maps allow you to organize different combinations of topics for different outputs and deliverables. You can reuse (repurpose) the same set of topics in different maps, for example to publish HTML from one map and a PDF book from another map. Each of these maps would refer to the same topics but reflect a different hierarchy.

You can also reuse individual topics in different maps, combining them with topics from different topic sets for different purposes, perhaps to reflect different user goals. For example, for a product one DITA map might assemble the topics for installation information, another might assemble the topics for the administration information. You can also combine different maps for different topic sets within a single map, so for example you could use one overall map to assemble all the topics for the different user goals of a product.

Obviously you need a tool to publish the output using the maps, and you can use the DITA Open Toolkit or other tools to do this. When you process a map to produce formatted output, the hierarchy becomes, for example, the nested sections in a book or the navigation tree in a help system or website.

A DITA map is defined by the map element in an XML file with the suffix `.ditamap`.

The following example illustrates a simple DITA map:

```
<?xml version="1.0" encoding="utf-8"?>

<map title="Example Map">
  <topicref navtitle="Parent concept" href="concepts/parentconcept.dita"
    type="concept"></topicref>
  <topicref navtitle="Concept A" href="concepts/concepta.dita"
    type="concept"></topicref>
  <topicref navtitle="Concept B" href="concepts/conceptb.dita"
    type="concept"></topicref>
</map>
```

The basic element of a DITA map is the `<topicref>` element, which can point to a DITA topic or another DITA map. The nesting of `<topicref>` elements defines the hierarchy. The attributes of `<topicref>` have the following purposes:

- `href`. A pointer to the DITA topic or map. The value of the `href` attribute is relative to the folder that contains the DITA map. It is good practice to store task, concept and reference topics in a subfolder of the same name, as shown in the examples.
- `navtitle` (optional). Specifies a title to be used in navigation panes or online contents.

- `type` (optional). Specifies the type of the topic. If the `navtitle` and `type` attributes are not specified, they can be taken from the target topic in the DITA publishing process.

DITA maps are also used to define the linking between topics, which is preferable to defining links within topics, as it allows you to change your linking within a map without having to change any topic files.

When you use a map to publish documentation, links between parent and child topic are added automatically, depending on the hierarchy defined in the map. So, for the example shown above, the following links might be generated in the output:

In parent concept	*In concept A*	*In concept B*
Concept A Short description	*Parent topic:* Parent concept	*Parent topic:* Parent concept
Concept B Short description		

However, you can also define links between topics explicitly using the `collection-type` attribute to show sequence or family relationships between sibling topics. You can set the attribute on the `<map>`, `<topicref>`, `<topicgroup>` and other elements in a DITA map. For example, the following shows a DITA map that uses the `collection-type` attribute on the `<topicgroup>` element.

```
<topicgroup collection-type="family">
<topicref navtitle="Concept A" href="concepts/concepta.dita"
type="concept"></topicref>
<topicref navtitle="Concept B" href="concepts/conceptb.dita"
type="concept"></topicref>
<topicref navtitle="Concept C" href="concepts/conceptc.dita"
type="concept"></topicref>
</topicgroup>

<topicgroup collection-type="sequence">
<topicref navtitle="Task A" href="tasks/taska.dita" type="task"></topicref>
<topicref navtitle="Task B" href="tasks/taskb.dita" type="task"></topicref>
<topicref navtitle="Task C" href="tasks/taskc.dita" type="task"></topicref>
</topicgroup>
```

This is how links might be generated for the family relationship:

In concept A	*In concept B*	*In concept C*
Related concepts	Related concepts	Related concepts
Concept B	Concept A	Concept A
Concept C	Concept C	Concept B

When you define a sequence relationship, next and previous links are added to the topics in the sequence (HTML-based output only):

In task A	*In task B*	*In task C*
Next topic:	*Previous topic:*	*Previous topic:*
Task B	Task A	Task B
	Next topic:	
	Task C	

You can also use the <reltable> element in a DITA map to define relationships between topics. The following is an example of a relationship table:

```
<?xml version="1.0" encoding="utf-8"?>
<map>
...
<reltable>
  <relheader>
    <relcolspec type="concept"/>
    <relcolspec type="task"/>
    <relcolspec type="reference"/>
  </relheader>
  <relrow>
    <relcell>
      <topicref navtitle="Concept A" href="concepts/concepta.dita"/>
    </relcell>
    <relcell>
      <topicref navtitle="Task A" href="tasks/taska.dita"/>
      <topicref navtitle="Task B" href="tasks/taskb.dita"/>
    </relcell>
    <relcell>
      <topicref navtitle="Reference A" href="ref/refa.dita"/>
    </relcell>
  </relrow>
```

```
    <relrow>
      <relcell>
        <topicref navtitle="Concept B" href="concepts/conceptb.dita"/>
      </relcell>
      <relcell collection-type="family">
        <topicref navtitle="Task B" href="tasks/taskb.dita"/>
        <topicref navtitle="Task C" href="tasks/taskd.dita" linking="targetonly"/>
      </relcell>
      <relcell>
        <topicref navtitle="Reference B" href="ref/refb.dita"/>
      </relcell>
    </relrow>
  </reltable>
  ...
  </map>
```

The markup in this example corresponds to a table like this:

Concept	*Task*	*Reference*	*Notes*
Concept A	Task A Task B	Reference A	In a row, each topic links to topics in other cells, but not to topics in the same cell...
Concept B	Task B Task C	Reference B	...unless the topics are siblings.

Links between topics whose relationships are defined in relationship tables appear as related links at the end of those topics after topics are processed. The following table shows some of the links that would be generated.

In task A	*In task B*	*In task C*

Related concepts
`Concept A`
	Related concepts
	`Concept A`
	`Concept B`
		No links, due to the `targetonly` value on the `linking` attribute of the `<topicref>` element.

Related reference `Reference A`	*Related tasks* `Task C`	You can also set the `linking` attribute on the `<relcell>` element.

	Related reference
	`Reference A`
	`Reference B`	

You will have gathered that DITA maps, and particularly relationship tables, can become quite complicated. A complete description of DITA maps is beyond the scope of this book; you can refer to the documentation available at `dita.xml.org/standard`, for example, for more information. Luckily there are modelling tools available that you can use to generate DITA maps, and some editors make coding of DITA maps relatively easy.

You might notice a similarity between DITA maps and the topic maps specification, an ISO standard for describing knowledge structures and associating them with information resources. The DITA maps concept does borrow many ideas from topic maps, but is more concerned with the practical application to technical documentation than topic maps, which provide a more general reference for organizing and managing knowledge.

DITA specialization

There are times when the elements in the standard DITA language specification are unsuitable for an organization's documentation projects. For example, to provide documentation for a particular programming language, specific elements might be required, or there may be a requirement for additional topic types, say for troubleshooting information. To extend an implementation of DITA, you can use *specialization* to create new elements as required.

Two types of specialization are covered by the DITA specification:

- *Structural specialization*, which is used to develop new information types
- *Domain specialization*, which is used to develop domain-specific vocabularies.

For *structural specialization*, new topic types are based on the generic topic type or the types based on it. For example, in DITA Version 1.1 you could create a Glossary topic type as a specialization of the Concept topic type, or a Troubleshooting topic type based on the Task topic type. (DITA version 1.2 introduces improved glossary markup).

New topic types inherit processing rules and other properties from existing types. You do not lose compatibility, so you can still share your DITA files and use the same XSLT stylesheets and tools to work with them.

You use *domain specialization* to define domain-specific vocabularies. DITA already has a number of domain specializations, including user-interface, software, programming and typography domains, but you can define more, for example to define elements suitable for a particular programming language.

You specify the names of the domains in use by the current document type on the `domains` attribute of the root element for the topic type. The following is an example of a task with three domains:

```
<task id="mytask" class="- topic/topic task/task " domains="(topic ui-d)
(topic sw-d) (topic pr-d jav-d)"> ... </task>
```

In this example, the task allows the use of tags for describing user interfaces (`ui-d`), software (`sw-d`), and also Java programming (`jav-d`). The Java programming domain is specialized from the ancestor programming domain (`pr-d`). The example also shows the ancestry of the task topic type on the `class` attribute.

A description of how to create specializations is beyond the scope of this book. For more information, see the *DITA Architectural Specification.*

DITA and reuse

In DITA you can reuse content in a number of ways:

- Reuse elements within and between topics
- Reuse topics in multiple places within one or more maps
- Reuse maps within maps
- Filter content for reuse in different contexts
- Reuse other XML content within DITA.

Apart from reusing content, you can reuse design by creating specialized information types, or domains that achieve reuse through inheritance of properties.

This reuse eliminates information redundancy and produces content that is more consistent and easier to maintain, thereby saving time and money. Furthermore, material is easier to share between products, groups and even organizations.

Figure 4.1 on page 79 illustrates various ways of reusing content in DITA.

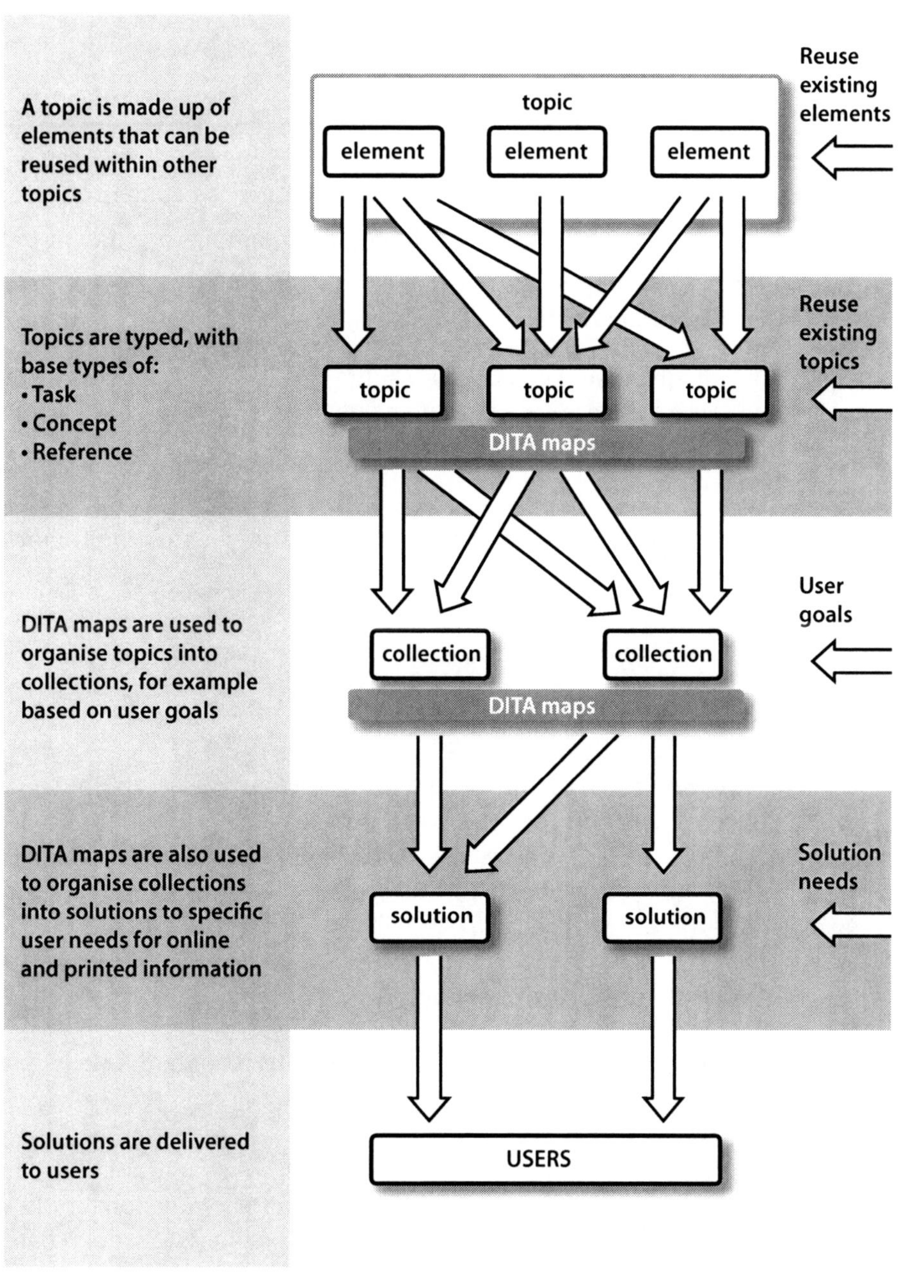
A topic is made up of elements that can be reused within other topics
topic
element
element
element
Reuse existing elements
Topics are typed, with base types of:
• Task
• Concept
• Reference
topic
topic
topic
DITA maps
Reuse existing topics
DITA maps are used to organise topics into collections, for example based on user goals
collection
collection
DITA maps
User goals
DITA maps are also used to organise collections into solutions to specific user needs for online and printed information
solution
solution
Solution needs
Solutions are delivered to users
USERS

Figure 4.1 Authoring and reuse with DITA (reproduced by kind permission of Ian Larner)

Reusing elements within topics

In DITA you can use content referencing to reuse elements from the same file or from different files. To do so, you use the `conref` attribute, which has the following syntax:

```
conref ="file.dita#topicid/elementid"
```

where:

- *`file`* is the filename of the topic containing the content to be reused
- *`topicid`* is the identifier of the topic containing the content to be reused
- *`elementid`* is the value of the `id` attribute of the element to be reused.

You can only use the `conref` attribute to reference elements of the same type. Using the `conref` attribute, you can for example reuse whole topics by referencing a topic id, paragraphs, or just single words if you use content referencing of phrase (`<ph>`) elements.

As an example, you could have a paragraph such as the following in a Concept topic:

```
<p>You can use the <ph conref="..\entities.dita#entities/editorprod"/> editor
with the <ph conref="..\entities.dita#entities/cmsprod"/> CMS as part of your
single-sourcing XML publishing solution.</p>
```

The `entities.dita` file would be stored in the same folder as your Concept topic and the referenced phrase elements might be as follows:

```
<ph id="editorprod">XOmatic</ph>
<ph id="cmsprod">XContent</ph>
```

When formatted, this paragraph would become:

> You can use the XOmatic editor with the XContent CMS as part of your single-sourcing XML publishing solution.

Common files such as the `entities.dita` file example are useful for reusing phrase-level content, and provide a similar functionality to general entities. They are therefore useful for managing product names and other volatile content.

As part of your reuse strategy, you should develop standards for file names, element ids and ownership of reused content. An alternative to reusing elements in a paragraph would be to produce alternative paragraphs and use metadata to control their reuse.

Reusing topics within maps

With DITA maps you can reuse the same set of topics in different maps for different output formats, as well as for different purposes.

In maps, you can also use the `conref` attribute to refer to a topic in a different map or in the same map. At publishing time, this includes not only the referenced topic, but also its children. For example, the following would reuse a topic with the id value of `financial` from the map file `verticals.ditamap`:

```
<topicref conref="verticals.ditamap#financial">
```

If you need to reuse a topic in several places within a map, you can do the following:

```
<topicref id="accounts" href="concepts\accounts.dita"
  navtitle="About Accounts" type="concept">
...
<topicref conref="#accounts">
```

By using DITA maps to manage links between topics, rather than hard-coding links in topics, you make topics more reusable.

Reusing maps

Apart from reusing topics in a DITA map, you can also reuse whole maps. This is useful if you have an existing map and you want to extend it for a particular purpose, or when you want to use the same set of topics in different publications. For example, in an overall map you can include another map as follows:

```
<topicref format="ditamap" href="verticals.ditamap"/>
```

Filtering content for reuse

In technical publications it is common to identify conditional content so that you can publish different versions of a document from a single source file. DITA uses the following metadata attributes in maps or topics to flag conditional content for various contexts:

- `product`: the product to which the content applies.
- `platform`: the operating system or environment to which the content applies.
- `audience`: the users to whom the content applies, for example, a user role such as ‘administrator’, or an experience level such as ‘beginner’ or ‘advanced’.
- `rev`: the revision level to which the content applies.

- `otherprops`: other contexts.

For example, in a Task topic you might identify conditional elements as follows:

```
<step platform="Windows"> <cmd>Log in with Administrator
  privileges.</cmd></step>
<step platform="UNIX"><cmd>Log in as root.</cmd></step>
```

You filter conditional content using a DITA filter file (filetype `.ditaval`), in which the `<prop>` element specifies what to include or exclude based on the values of the `product`, `platform` and other metadata attributes in your source files. For example:

```
<val>
   <prop att="platform" val="Windows" action="include" />
   <prop att="platform" val="UNIX" action="exclude" />
</val>
```

When you publish the documentation, you specify which filter file to use. For the filter file shown in the example, the conditional content for the Windows platform is included. In this way you can control the reuse of topics, maps and elements by using metadata to flag and filter content for specific contexts.

Reusing content from other sources

It is quite straightforward to reuse external XML content in your DITA projects. Often in software projects XML is used for definitions of graphical user interfaces items, for error messages and for software property files. By using XSLT processing or other programming, you can transform content from such files so that you can use it in your DITA source files.

For example, to ensure consistency with the software application you are documenting, you could transform an XML file containing GUI string names to a topic file containing phrases that you can reuse in your source files using the content referencing mechanism.

Tools available for working with DITA

The DITA Open Toolkit is an open source implementation of the OASIS specification for DITA that you can download from `http://sourceforge.net/projects/dita-ot/`. The Toolkit contains the DTDs and Schemas for DITA, sample files, XSLT transforms and other tools for producing a variety of output formats. You can use it to generate XHTML, PDF, HTML Help, RTF and a number of other formats.

The toolkit can generates indexes, tables of content and cross-references, and apply the required formatting for particular output formats. It also filters conditional content according to settings in the `.ditaval` file.

Although you can use the toolkit out of the box, you can also customize or extend it to change the formatting of the output formats or perhaps produce different formats. The toolkit uses Ant, an open-source tool provided by the Apache Foundation, to control its processing when publishing output from source files. Ant system scripts define the standard processing for a range of outputs. When you build your own project documentation, you create an Ant user script that defines what you want to build and calls the system scripts to perform standard processing. By adjusting property settings in the user script, you can customize the build process, and by adding your own XSLT files you can extend the standard XSLT processing. For more information about using the toolkit, see the *DITA Open Toolkit User Guide.*

You can use the toolkit on its own, but it is also built into a number of tools that support DITA file processing. For example, the DITA edition of the XMetal editor integrates the toolkit into its authoring environment, as does the Arbortext editor and a number of other editors.

Support for DITA is available in FrameMaker 7.2 through an application pack for DITA. This allows you to import and export DITA source files using a structured application, with the associated template, read/write rules file and other files. The DITA EDD is based on the `ditabase` DTD or schema. However FrameMaker 7.2 does not provide full DITA support: for example, DITA maps and `conrefs` are not supported. These shortcomings are addressed with the DITA structured application supplied with FrameMaker 8.0 and later versions.

A number of content management systems (CMS) support DITA. For example, Sibersafe from Siberlogic, with the DITA Application Pack configured, provides seamless DITA content management functionality from within the FrameMaker user interface.

The advantages of DITA

DITA has many advantages for technical documentation: it is well suited to single-sourcing publishing, it is particularly good for the reuse and repurposing of content for a wide range of documentation deliverables, and of course, being an open standard, it is free.

DITA encourages best practices in information architecture through its topic-based architecture and enforcement of consistent information structure across topics. Consistency in the presentation of similar information helps users navigate the information more quickly, and consistently structured topics are easier to reuse in multiple deliverables.

DITA has a number of advantages for localization. It supports translation by providing the `translate` attribute, available on most elements, which indicates whether or not content should be translated. DITA's topic-orientation and reuse features make it easier to manage translation projects and promote cost saving. A further advantage in DITA 1.1 is the automatic alphabetization of indexes and glossaries into multiple languages.

DITA and DocBook compared

DocBook and DITA are both XML languages aimed at the technical documentation market, so it is worthwhile to compare them.

In DocBook, content is organized in terms of a book structure, with elements for chapters, parts, sections and so on. It was designed for a single, linear deliverable such as a book. With DITA, on the other hand, authors develop information as discrete topics that are assembled and transformed into any kind of deliverable, including printed books, though it is possibly best suited for topic-oriented content such as websites, online help and other online information. Nevertheless, DITA 1.1 introduced a Bookmap specialization of the DITA map that supports long books and contains numerous metadata elements for front matter and back matter.

DITA breaks information into a much more granular structure than DocBook, which means content reuse can be more granular. In particular, DITA's content reference mechanism provides a flexible reuse capability.

Because DocBook has been used for much longer than DITA, vendors have had the time to develop and refine tools that support it, although as DITA is adopted more, its tools support can only improve. It might currently be easier to exchange content with DocBook: because more people use DocBook, it is easier to trade and interchange content using DocBook.

It is arguably easier to learn the basics of DocBook than DITA. DocBook follows a familiar book paradigm, and although there are many elements in DocBook (some would argue too many), a technical communicator can easily relate to them. In DITA, on the other hand, you have to understand the information typing architecture as well as the actual markup available. You also need to know about object-oriented concepts to understand how DITA works in terms of inheritance of properties and specialization.

In DocBook, it is possible to extend the language through a customization layer, though this is often frowned on, as it may hinder tool support and interoperability. On the other hand, DITA is probably more supportive of extension, as specialization to create new information types for specific purposes requires less customization work, because markup and rules are inherited from an existing document type and overall interoperability is maintained. However, it can also be argued that DocBook, with its rich collection of elements, has less need of customization compared to DITA, where you probably need to develop some of your own information types.

S1000D

S1000D is an international specification for the production and procurement of technical publications. It originated in the early 1980s and was intended as a global standard similar to ATA 100, the civil airline standard for technical manuals written by aviation manufacturers and suppliers. S1000D was developed initially by the Association Européene des Constructeurs de Matériel Aérospatial (AECMA) as a means of managing maintenance documentation for military aircraft. Later, AECMA became the Aerospace and Defence Industries of Europe (ASD) and assumed responsibility for the governance of the S1000D specification, together with the Aerospace Industries Association of America (AIA). However, the actual development and maintenance of S1000D is controlled by the Technical Publications Specification Management Group (TPSMG) of ASD and its various workgroups.

Although S1000D was designed initially for aircraft maintenance documentation, it has also been used for aircrew-related technical documents and for other purposes. It has been applied mainly to military aerospace projects, although the specification covers land, sea and commercial equipment as well as aviation equipment.

The S1000D specification covers the planning, management, production, exchange, distribution and use of technical documentation. There have been a number of issues of the specification; earlier issues supported SGML only, while later ones provided XML DTDs and XML Schemas to define the markup language for the specification.

S1000D content is based on *data modules*, which are stand-alone chunks of information that can be coded in SGML or XML. This modularity promotes reuse of information, facilitating its management in a common source database (CSDB), and its delivery, both as traditional page-oriented manuals and electronic publications. There are a number of different types of data module, for which SGML and XML DTDs and XML Schemas (XSD files) are available, depending on the issue of the specification.

The website `www.s1000d.org` is the official resource for S1000D information. You can download various issues of the S1000D specification as PDF files, together with XML DTDs and schemas for each type of data module and example data modules that conform to the specification. The S1000D specification is a very large document, containing well over 2,500 pages. As of early 2010, the current version was Issue 4.0, for which only XSD files, not DTDs, are available. An Issue 4.0.1 patch is available, so new projects based on Issue 4 should use the files from this or any later patch.

Note that in this book the XML examples are valid for S1000D Issue 4.0 markup. You should be aware that there are substantial differences in the name and availability of elements and attributes between issues of the specification. Although Issue 4.0 is the latest version at the time of publishing of this book, many projects are using earlier versions of the specification.

Data modules

A data module is described in the S1000D specification as:

> *'A self-contained unit of data for the description, operation, identification of parts or maintenance of the product and its support equipment. The unit of data consists of an identification and status section and a contents section, and is produced in such a form that it can be input into and retrieved from a database using the data module code as the identifier.'*

As mentioned in the definition, a data module has two main parts:

- *The identification and status section.* This section comprises information that includes a unique data module code (DMC) for identifying the data module, together with metadata that is used for managing the data module, providing status information and so on.

- *The contents section.* This section comprises the actual contents of the data module, which, depending on the type of data module, may be the information seen by the user, support data management and reuse, or define the business rules for managing the documentation project.

Apart from these sections, a data module can also contain Resource Description Framework (RDF) and Dublin Core (DC) markup as an additional means of providing metadata for the data module.

Although data modules might contain a lengthy procedure or description, they can also contain a short boilerplate paragraph, a warning message and so on. This means that the number of data modules for a project can be considerable - a medium-size project might have thousands of data modules.

A set of sample data modules for the documentation of a bicycle is available from `www.s1000d.org`. Figure 4.4 on page 121 shows an abridged and annotated version of one these data modules, which describes the procedure for removing the rear wheel of a bicycle. In sections of this book, some of the markup from this data module is used to illustrate particular S1000D concepts.

The following sections describe data modules in detail.

Data module identification and status section

In a data module, the `<identAndStatusSection>` element contains all the elements necessary to identify and manage the data module. There are two parts: identification and status.

In the identification part, the `<dmAddress>` element contains the unique identification, together with additional information to support the data module. Child elements include:

- `<dmIdent>`, which has child elements containing the data module code `<dmCode>`, indicating the language in which the data module is written `<language>`, and the issue number of the data module `<issueInfo>`.
- `<dmAddressItems>`, which contains the issue date and the data module title.

The following is an example of an identification part:

```
<dmAddress>
  <dmIdent>
    <dmCode modelIdentCode="S1000DBIKE" systemDiffCode="AAA"
     systemCode="DA0" subSystemCode="2" subSubSystemCode="0"
     assyCode="00" disassyCode="00" disassyCodeVariant="AA"
     infoCode="520" infoCodeVariant="A" itemLocationCode="A">
    </dmCode>
    <language countryIsoCode="US" languageIsoCode="en"/>
    <issueInfo issueNumber="006" inWork="00"/>
  </dmIdent>
  <dmAddressItems>
    <issueDate year="2008" month="08" day="01"/>
    <dmTitle>
      <techName>Rear wheel</techName>
      <infoName>Remove procedures</infoName>
    </dmTitle>
  </dmAddressItems>
</dmAddress>
```

In the status part, the `<dmStatus>` element contains various metadata items. There can be many other optional child element, but the mandatory ones are as follows:

- `<security>`, which indicates the security classification of the data module and its illustrations, and can determine who can make changes to the data module.
- `<responsiblePartnerCompany>`, which contains the company or organization responsible for the data module. There can be multiple companies involved in a project, so the company responsible for the data module may be different from the originating company.
- `<originator>`, which indicates the originating company or organization responsible for the production of the data module.
- `<applicCrossRefTableRef>`, which contains a pointer to the applicability cross-reference table (ACT) data module used to declare product attributes that apply to the data module - see *Applicability* on page 103.
- `<applic>`, which contains applicability annotations. These can be used to apply conditional processing - see *Applicability* on page 103.
- `<brexDmRef>`, which contains a reference to the business rules exchange (BREX) data module that reflects the business rules that apply to the data module. For more information, see *The BREX data module* on page 100.

- <qualityAssurance>, which contains details of the status of the quality assurance (QA) process as required by the project, and indicating whether or not the data module has been verified.

The following is an example of a status part that contains only mandatory elements.

```
<dmStatus issueType="revised">
  <security securityClassification="01"
   commercialClassification="cc51" caveat="cv51"/>
  <responsiblePartnerCompany enterpriseCode="U8025">
    <enterpriseName>UK MoD</enterpriseName>
  </responsiblePartnerCompany>
  <originator enterpriseCode="U8025">
   <enterpriseName>UK MoD</enterpriseName>
  </originator>
  <applicCrossRefTableRef>
    <dmRef xlink:type="simple" xlink:actuate="onRequest"
     xlink:show="replace"
     xlink:href="URN:S1000D:DMC-S1000DBIKE-AAA-D00-00-00-00AA-00WA-D">
      <dmRefIdent>
        <dmCode modelIdentCode="S1000DBIKE" systemDiffCode="AAA"
        systemCode="D00" subSystemCode="0" subSubSystemCode="0"
        assyCode="00" disassyCode="00" disassyCodeVariant="AA"
        infoCode="00W" infoCodeVariant="A"
        itemLocationCode="D"></dmCode>
      </dmRefIdent>
    </dmRef>
  </applicCrossRefTableRef>
  <applic>
    <displayText>
      <simplePara>Mountain bicycle and (Mountain storm
       Mk1 or Brook trekker Mk9)</simplePara>
    </displayText>
    <!-- Applicability annotations not shown -->
  </applic>
```

```
    <brexDmRef>
      <dmRef xlink:type="simple" xlink:actuate="onRequest"
       xlink:show="replace"
       xlink:href="URN:S1000D:DMC-S1000DBIKE-AAA-D00-00-00-00AA-022A-
             D_007-00">
        <dmRefIdent>
          <dmCode modelIdentCode="S1000DBIKE" systemDiffCode="AAA"
           systemCode="D00" subSystemCode="0" subSubSystemCode="0"
           assyCode="00" disassyCode="00" disassyCodeVariant="AA"
           infoCode="022" infoCodeVariant="A" itemLocationCode="D"/>
          <issueInfo issueNumber="007" inWork="00"/>
        </dmRefIdent>
      </dmRef>
    </brexDmRef>
    <qualityAssurance>
      <firstVerification verificationType="tabtop"/>
    </qualityAssurance>
  </dmStatus>
```

Data module content section

In a data module the allowed markup in the <content> element depends on the type of the data module. Table 4.2 shows the types of data module available in Issue 4.0 of S1000D. There is an XSD file for each of these, and there are also XSD files for common elements and attributes, entities, a catalog and so on.

Table 4.2 Data module types in Issue 4.0 of S1000D

Data module type	***Information and usage***
Data modules that provide user information	
Crew/operator	Operational checklist procedures and related descriptive information typically used in aircrew flight reference cards and flight manuals.
Descriptive	General purpose descriptive text. Useful for legacy data because of its flexible structure.
Fault isolation	Information for fault reporting and fault isolation.
Illustrated parts data	Information for a parts catalogue in which each data module includes a labelled parts illustration with a corresponding parts list.
Maintenance checklists and inspections	Procedural tasks that must be presented with the maintenance data.

Table 4.2 Data module types in Issue 4.0 of S1000D (continued)

Data module type	***Information and usage***
Maintenance planning	Information for planning of maintenance.
Procedural	Information describing step-by-step procedures or instructions.
Process	Interactive information; mainly containing a type of scripting language executed by a logic engine to create interactive information. Used for example to capture user input and deliver maintenance information dynamically depending on equipment status.
Learning	Technical training information.
Wiring data	Wire, harness, equipment and standard parts data. This information can be used to generate wiring and schematic diagrams.
Wiring data description	Wiring element field descriptions for each relevant element in a wiring data module.
Data modules that support data management and reuse	
Applicability cross-reference table (ACT)	Declarations of product attributes - see *Applicability* on page 103.
Conditional cross-reference table (CCT)	Declarations of the conditions that can affect applicability of data - see *Applicability* on page 103.
Container	Associations of several alternate data modules. For example, the container module can associate a number of similar maintenance tasks that may differ based on a particular configuration. See *Reuse through the container data module* on page 114.
Product cross-reference table (PCT)	Definitions of product instances and their associations with product attributes and conditions - see *Applicability* on page 103.

Table 4.2 Data module types in Issue 4.0 of S1000D (continued)

Data module type	***Information and usage***
Technical information repository	Information about, for example, functional items that can be cross-referenced from data modules. See *Reuse through the technical information repository data module* on page 114.
Other data modules	
Business rules information (BREX)	Project rules to be followed during data creation, exchange, and delivery. Essentially a configuration file that defines the markup allowed and how it is used for a project. For more information, see *Business rules* on page 98.

Many of the data module types in the table, mainly those providing user information, are said to correspond to *information types*, according to the S1000D specification.

The available data module types vary with the issue of the specification, and the name of the data module as used in the specification may also vary. Several entries in the table refer to the product, which in the S1000D specification is defined as '*any platform, system or equipment (air, sea, land vehicle, equipment or facilities, civil or military)*'.

Common constructs in data modules

There are specific elements for the different data module types, but there are many common constructs within data modules, as shown in Table 4.3.

The XSD files for S1000D specify the elements available for S1000D markup. However, it is important to realise that a project can determine whether particular elements are not used through the use of business rules.

Table 4.3 Common constructs shared by data modules

Common construct	*Elements*
Change marking	Elements and attributes for indicating what has changed and the reason for the changes: ■ The `<reasonForUpdate>` element, which contains a reason for updating a data module. ■ The `<changeInline>` element, used for changes within certain elements (for example, paragraphs and titles). ■ The following attributes, used on `<changeInline>` or on other elements: – `changeMark`, which indicates a change bar or mark to be used next to changed content – `changeType`, which indicates the type of change with values like add, modify, and insert – `reasonForUpdateRefIds`, which references a `<reasonForUpdate>` element.
Referencing	Elements for cross-referencing: ■ To other places within the same data module, `<internalRef>` ■ To other data modules, `<dmRef>` ■ To publication modules, `<pmRef>` ■ To non-S1000D publications or documents, `<externalPubRef>`.
Lists	Elements for: ■ Sequential (ordered) lists, `<sequentialList>` ■ Random lists, `<randomList>`, which can be simple or unordered ■ Definition lists, `<definitionList>`, for terms and definitions.
Caption groups	The elements `<caption>`, used to represent a single illuminated warning light, push button, annunciator or display caption, and `<captionGroup>`, which represents groups of those items.
Titles	The `<title>` element for the title of a paragraph, step, table or figure and so on.

Table 4.3 Common constructs shared by data modules (continued)

Common construct	*Elements*
Tables	The `<table>` element and its child elements for two types of table: ■ Formal tables, consisting of four parts: – The table title, `<title>` – The table head, `<thead>` – The optional table footer, `<tfoot>` – The table body, `<tbody>`. Each of these parts contains the rows and entries that comprise the table, `<row>` and `<entry>`. ■ Informal tables, which are short, simple tables, without a table title, table head and table footer.
Figures and foldouts	The `<figure>` element, which comprises a title, one or more graphics and an optional legend, and the `<foldout>` element, which contains a figure that is larger than the default size.
Hotspots	The `<hotspot>` element, which contains the definition of graphical regions in an illustration, together with information needed to navigate between graphical objects or between graphical objects and text.
Preliminary requirements and requirements after job completion	The `<preliminaryRqmts>` element, which contains, for example, actions to be done, conditions to be satisfied, information or equipment required before a main procedure is performed. The `<closeRqmts>` element, which contains any actions to be done, or conditions to be satisfied, after the main procedure is complete.
Text elements	Elements for paragraphs: `<para>`, `<warningAndCautionPara>`, `<attentionListItemPara>`, `<notePara>`, `<simplePara>` and their child elements. Also the `<indexFlag>` element, used when generating an index.
Controlled content	The attributes `authorityName` and `authorityDocument`, which are used to indicate controlled content.
Common information	The `<commonInfo>` element, used to provide data to the user that applies to an entire data module.

In the schemas supplied with the S1000D specification, there are three classes of attributes, all of which have predefined allowable sets of values:

- *Class 1 - Specific.* These attributes represent project-specific information, and to some extent can be tailored to meet project requirements.
- *Class 2 - Generic.* These attributes represent characteristics and properties of general interest and relevance, and must be applied strictly in accordance with the specification, which describes their intended use.
- *Class 3 - Public.* These attributes are incorporated or inherited due to reuse of structures, elements and attributes from contexts external to S1000D. An example is the `align` attribute of the Continuous Acquisition and Lifecycle Support (CALS) table model.

For each Class 1 attribute, a project can decide (in the business rules) to use only a subset of the predefined values, and also determine project-specific interpretations of attribute values.

Every data module has a unique data module code (DMC) that allows data to be identified and managed in a common source database (CSDB). The DMC also indicates how subcomponents of a piece of equipment relate to larger components.

The structure of the DMC is reflected in the attributes of the `<dmCode>` element. The following example represents the DMC (S1000DBIKE-AAA-DA0-20-00-00AA-520A-A) for the bicycle wheel removal data module:

```
<dmCode modelIdentCode="S1000DBIKE" systemDiffCode="AAA"
      systemCode="DA0" subSystemCode="2" subSubSystemCode="0"
      assyCode="00" disassyCode="00" disassyCodeVariant="AA"
      infoCode="520" infoCodeVariant="A" itemLocationCode="A">
</dmCode>
```

The following list describes the various parts of the DMC:

- *Model identification code* (`modelIdentCode`). Indicates the product or system to which the content is applicable. Projects must apply to the NATO Maintenance and Supply Agency (NAMSA) for the allocation of the model identification code.
- *System difference code* (`systemDiffCode`). Indicates alternative versions of the system and subsystem/sub-subsystem identified by

the standard numbering system. Different codes are used to identify information that is specific to a unique configuration of a product.

- *Standard numbering system (SNS).* Indicates the system (`systemCode`), subsystem/sub-subsystem (`subSystemCode`, `subSubSystemCode`) and individual units or assemblies (`assyCode`) to which a data module applies. Each project or organization must define the SNS structure that is used in its business rules. You can use the relevant SNS for the equipment type, (for example, air vehicle, navigation system), as defined in the S1000D specification, or alternatively you can allocate a project-specific SNS. For the example, the SNS for the bicycle example above is `DA0-20-00`.
- *Disassembly code* (`disassyCode`). Identifies the breakdown condition of an assembly to which maintenance information applies.
- *Disassembly code variant* (`disassyCodeVariant`). Distinguishes between two or more data modules that share the same number.
- *Information code* (`infoCode`). Identifies the purpose of a data module. A three-character code taken from the predefined list of information codes in the S1000D specification. In the example, code `520` is for 'remove procedure'.
- *Information code variant* (`infoCodeVariant`). Distinguishes between generic information codes, and identifies any variation in the activity defined by the information code.
- *Item location code* (`itemLocationCode`). Identifies the situation to which the information is applicable – for example, where the maintenance task for a product is carried out.

Learning data modules have the following additional parts in their DMC:

- *Learn code.* Identifies the type of human performance technology or training information in the content of the data module. The learn code is used for projects that are Shareable Content Object Reference Model (SCORM) conformant or that want to use the functionality brought about by the learn code.
- *Learn event code.* Identifies the type of learning information, if a learn code is used.

The minimum number of characters allowed in the whole DMC is 17 and the maximum is 41 (or 37 if the learn code and learn event code are not included).

The DMC is used as a data module's identifier in the `<dmRef>` element for linking between data modules. The DMC is also used as the basis for the data module's file name, for example,

```
DMC-S1000DBIKE-AAA-DA0-20-00-00AA-520A-A_006-00_EN-US.xml
```

...the file name for the sample in this section.

The S1000D documentation process

The workflow for the S1000D documentation process is shown in Figure 4.2.

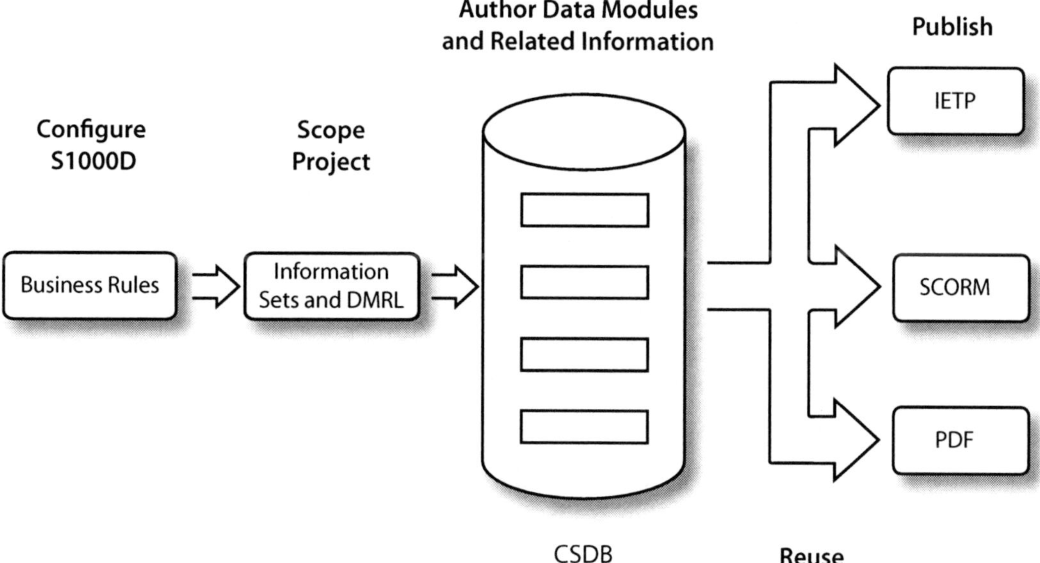

Figure 4.2 S1000D documentation process

It consists of the following steps:

1. Establish the business rules to customize S1000D for the project.
2. Agree the information sets to determine the scope and depth of the project.
3. Generate the data module requirements list (DMRL), which is the list of data modules for the project.
4. Generate the data modules and related information, which are stored in the CSDB. Add applicability annotations as appropriate to allow conditional processing.

- For each customer, agree the publications, deliverable media and front matter. Publications can, for example, be page-oriented, interactive electronic technical publications (IETP), SCORM publications or other formats.
- Generate the publications, including front matter, using publication modules and SCORM content packages to organize the data modules, XSLT stylesheets and XSL-FO to apply the appropriate visual presentation and formatting.

The following sections discuss each of these steps.

According to the S1000D specification:

> *'Business rules are decisions that are made by a project or an organization on how to implement S1000D. Business rules cover all aspects of S1000D and are not limited to authoring or illustrating. They can also address issues that are not defined in S1000D, such as rules related to how S1000D interfaces with other standards, specifications and business processes that are related to its implementation.'*

Business rules cover decisions about the creation, management, interchange and delivery of technical content, data integrity, 'legacy' data conversion and use of associated standards. There are ten categories of decision:

- *General business rules.* All decisions made by a project or an organization that are not covered by any of the specific business rule categories.
- *Product definition business rules.* The data module coding strategy related to the physical or functional breakdown of the product. These include the definition of the model identification codes and SNS to be used.
- *Maintenance philosophy and concepts of operation business rules.* The types of information that a project or an organization requires. This category covers, for example, the information sets and information codes used by the project.
- *Security business rules.* Security issues, including security classifications, copyright markings, use or disclosure restrictions, destruction instructions and any other data restrictions.

- *Business process business rules.* How technical publications development is coordinated with other disciplines within an organization, or within the project level at that organization, or the project as a whole.
- *Data creation business rules.* Information about the creation of textual data, graphics, 3D content and multimedia objects.
- *Data exchange business rules.* How data must be exchanged between partners and customers.
- *Data integrity and management business rules.* How data integrity within the CSDB is enforced.
- *Legacy data conversion, management and handling business rules.* Rules for converting non-S1000D data to data module format, which will vary from project to project, depending on the format of the source data and how the target data modules must be treated. To some extent these rules can be considered as being outside the scope of S1000D.
- *Data output business rules.* The output formats for S1000D data, which can include page-oriented formats, interactive electronic technical publication (IETP) formats, multimedia formats and SCORM formats.

This means that many business rule decisions are covered in the S1000D specification - over 600 in fact. In this chapter you will see some examples of where business rules are used, for example in applicability and in the definition of a project SNS.

Business rules that govern what can and cannot appear in a project's content must be enforced programmatically. The BREX data module is used to enforce such business rule decisions, as it contains XPath statements that can be used to check authored content (either during or after authoring) to ensure that the content contains only values or constructs specified in the project's business rules. Other business rules that are not enforced programmatically are specified in a business rules document and must be enforced by project or organization policy.

Business rules can be layered. The project BREX data module can be based on and refer to an organization BREX data module, which itself can be based on and refer to a set of even more general rules defined for a higher-level organization (for example, the civil aviation industry), or for national business rules. However, a BREX data module can only contain restrictions to the rules defined by the BREX data module on which it is based. At the top level are the S1000D default BREX data module and the XML schemas, which reflect the rules imposed by the specification itself.

Writing business rules is not particularly easy, neither is it easy to know, when authoring a data module, that you are conforming to the business rules. Fortunately software is available to help with these tasks.

The BREX data module

The business rules for a project are contained in a BREX (business rules exchange) data module, which can contain:

- Specifications of SNSs that apply to the project
- Specifications of elements and attributes that must or must not be applied to CSDB objects generated for the project (that is, the data modules and other XML documents)
- Definitions of which values are allowed/used for specified elements and/or attributes, and how these values are interpreted
- Descriptions of the purpose of markup elements and attributes.

Every data module refers to a BREX data module for the business rules that apply to the data module. This can be either a BREX data module created for a project or the default BREX data module, if it is required to implement S1000D in full. You can download the default BREX data module from `www.s1000d.org`.

The optional `<snsRules>` element and its child elements provide descriptions of one or more SNS systems that apply to the project.

Rules about the project-specific use of elements and attributes for the various schema used in the project are specified in the `<structureObjectRuleGroup>` element. The `<structureObjectRule>` element contains a particular rule for an element or attribute; it has child objects as follows:

- The `<objectPath>` element, which contains an XPath address that defines where the element or attribute occurs in the schema structure. The attribute `allowedObjectFlag` specifies whether the element or attribute should be included (1) or excluded (0).
- The `<objectUse>` element, which gives a description of the purpose of the element or attribute.
- The optional `<objectValue>` element, which gives values that are applicable to an element or attribute, together with short descriptions to explain the meaning of the value.

The following simple example shows how the attribute `learnCode` is mandated in the DMC of a data module:

```
<structureObjectRule>
  <objectPath allowedObjectFlag="1">//dmIdent/dmCode/@learnCode
  </objectPath>
  <objectUse>Learn code must always be included in the data module
   Identifications
  </objectUse>
</structureObjectRule>
```

The following more complex example shows how the use of either the `<systemBreakdownCode>` element or the `<functionalItemCode>` element (in the status section) is mandated - or rather, how their absence is prohibited.

```
<structureObjectRule>
  <objectPath allowedObjectFlag="0">not(//dmStatus/systemBreakdownCode
  or //dmStatus/systemBreakdownCode)
  </objectPath>
  <objectUse>Responsible partner company code in ICN must be uppercase A
  </objectUse>
</structureObjectRule>
```

Business rules specified in a BREX data module must define the context to which they are related, which can be the entire project or just one of the schemas applied within the project. Rules given for a specific context override the rules specified for all contexts. The `<contextRules>` element gives the rules for different contexts within a project, as illustrated in the following example:

```
<!-- A structure rule specific to the comment schema -->
<contextRules rulesContext="http://www.s1000d.org/S1000D_4-
0/xml_schema/comment/commentSchema.xsd">
<structureObjectRuleGroup>
  <structureObjectRule>
    <objectPath>//comment</objectPath>
    <objectUse>The comment object is not applied to these
    projects</objectUse>
  </structureObjectRule>
</structureObjectRuleGroup>
</contextRules>
<!-- Rules applicable to all schemas -->
<contextRules>
  <structureObjectRuleGroup>
<!-- Other rules -->
  </structureObjectRuleGroup>
</contextRules>
```

Information sets

In the S1000D specification *information sets* define the purpose, scope and depth of the technical information to be produced for operation and maintenance of the product. Information sets are used to produce the list of data modules for a project (DMRL) and also in defining the publications delivered for the project.

The specification describes the content of a number of common information sets, for example, crew/operator information, description and operation, maintenance procedures and fault isolation, to name just a few of the 25 listed in the Issue 4.0 specification. (There are also air-specific information sets and land/sea-specific information sets.) For each information set, the specification prescribes the information (topics, instructions, procedures, lists and so on) that must be provided in the data modules for the information set. In many cases, it also shows how the DMC for data modules should be coded, for example to include the information code appropriate for the type of information in the data module.

Information sets relate to publications, in that an information set can be a subset of or equal to an information set, but it can also be a superset of several information sets or parts of them.

Data module requirements list

The data module requirements list (DMRL) is an XML file that lists the data modules required for a project or part of a project. For a project there can be a single complete DMRL or a number of partial DMRLs, which is useful, for example, when partner companies are involved in the project.

DMRLs are defined at the start of a project and are managed in the CSDB. The DMRL aids such tasks as resource planning, production and configuration control, and can identify opportunities for reuse. A similar list, the CSDB status list, can be used to give the status of a CSDB for a project.

A DMRL has an identification and status section, which contains a data module list code (similar to a DMC) and other information to identify the DMRL uniquely. The body of the DMRL is contained in the `<dmlContent>` element, which contains a `<dmEntry>` element for each data module.

Within the <dmEntry> element, <dmRef> elements reference the data modules. The following example shows the structure of a DMRL:

```
<dmlContent>
  <dmEntry>
    <dmRef>
   <!--markup omitted -->
    </dmRef>
  </dmEntry>
</dmlContent>
```

Applicability

In S1000D information can be identified as being *applicable* to a specific condition or product attribute. Applicability information allows you to specify which information is appropriate for particular situations, enabling conditional processing of information to produce documentation tailored for different configurations.

Conditions can be technical conditions to do with the configuration of the product, for example service bulletins or modifications, or operational and environmental conditions, such as the availability of tools, regulatory rules, temperature, wind speed. Conditions are specified in the conditional cross-reference table (CCT).

Product attributes are properties of a product, such as model, series and serial number, which are typically set at the time of manufacture of a product instance and do not change during the service life of the instance. A *product instance* is an actual physical product, for example an Acme Moto-X Series 6 motorbike, which might have a serial number such as 2C080754. Product attributes are specified in the applicability cross-reference table (ACT), while product instances are specified in the product cross-reference table (PCT).

Applicability information is used to generate a filtered view of documentation for publication or display in a viewer. Applicability annotations in data modules are used to filter content, and only show the information that is appropriate for a particular condition, product attribute or combination of conditions and attributes. For example, information relevant only to a particular model of a product might be displayed, or for a particular condition, and so on. The filtering can apply to a whole data module or to particular content within the data module, as described in *Global applicability and inline applicability* on page 106.

For applicability filtering to work, data modules normally reference the ACT, which is used whenever displaying data or filtering for publication. As the ACT references the CCT and PCT, all other data modules are able to access all declarations of product attributes and conditions, as well as actual values for product instances.

Applicability branches

In data modules, applicability annotations are contained in the <applic> element within the identification section. There are three types of annotation, which are referred to as *branches* in the S1000D specification:

- Human-readable applicability annotation
- Computable applicability annotation
- Process data module applicability expression.

A *human readable applicability annotation* consists of the <displayText> element, which provides information that will be displayed in output. This element can be used on its own or with either of the other two types of annotation. For example:

```
<applic>
  <displayText>
    <simplePara>Motor Bike, either:</simplePara>
    <simplePara>Moto-X Series 6 or Yamazuki 350</simplePara>
  </displayText>
</applic>
```

A *computable applicability annotation* consists of the <assert> and <evaluate> elements, and is used to filter information according to conditions and product attributes. The <assert> element specifies the product attribute or condition to test and a set of values and/or ranges to test against. The result of the test is the Boolean value `true` or `false`. The <evaluate> element groups tests and provides the logical operation to apply to the test results. To test for the attribute or condition, reference is made to the ACT for the data module, which in turn references the CCT and PCT.

The following example illustrates an applicability annotation in which two product attributes are being tested and both must have the Boolean value `true` for the entire applicability to result in the value `true`. The model attribute must be 'Yamazuki' and the version attribute must be '350', for the relevant text 'For Yamazuki 350' to be displayed. If either the model or version test results in the value `false`, the entire applicability results in the value `false`.

```
<applic>
  <displayText>
    <simplePara>For Yamazuki 350 only</simplePara>
  </displayText>
  <evaluate andOr="and">
    <assert applicPropertyIdent="model"
            applicPropertyType="prodattr"
            applicPropertyValues="Yamazuki"/>
    <assert applicPropertyIdent="version"
            applicPropertyType="prodattr"
            applicPropertyValues="350"/>
  </evaluate>
</applic>
```

A *process data module applicability expression* consists of the element `<expression>`, which provides an applicability expression for the logic engine used with the process data module. This branch is used only within process data modules. The branches of the applicability model that are used by a project or organization are determined in the business rules.

Apart from its use in filtering of information, applicability can be used in a static view in which *all* the applicability information is presented on output, such as a procedure in which steps that vary for different models of a product might be presented. In the output, the steps would be flagged to show, for example, which are applicable for each model:

1 Locate the filter widget assembly.

2 *For Yamazuki 350 only:* Disengage the security catch.

3 Remove the filter element.

If just static applicability is required, only the human-readable branch of the applicability annotation is used within data modules, and in this case the ACT, CCT and PCT data modules are not needed. On the other hand, if applicability filtering is required, the full applicability annotation, including the ACT, CCT and PCT data modules, are used. For each project, therefore, you must decide to which level applicability is to be implemented: this is determined in the business rules.

Global applicability and inline applicability

Applicability annotations in the identification and status section of the data module apply to the whole data module. This global applicability means that you can filter whole data modules for relevance: for example, you could include only data modules relevant to a particular product model in a publication.

However, you can also add inline applicability information in the content section of data module that references annotations in the identification and status section. This allows you to filter parts of a data module.

For inline applicability, annotations are defined using the `<applic>` element within the `<referencedApplicGroup>` element. In the content section, the applicability annotations are applied by using the `applicRefId` attribute on the relevant element. This is illustrated in the following example:

```
<identAndStatusSection>
<!--markup omitted -->
  <referencedApplicGroup>
    <applic id="app-0001">
      <displayText>
        <simplePara>Moto-X Series 6</simplePara>
      </displayText>
      <evaluate andOr="and">
        <assert applicPropertyIdent="model"
        applicPropertyType="prodattr" applicPropertyValues="Moto-X"/>
        <assert applicPropertyIdent="series"
        applicPropertyType="prodattr" applicPropertyValues="Series 6"/>
      </evaluate>
    </applic>
  </referencedApplicGroup>
<!--markup omitted -->
</identAndStatusSection>
<!--markup omitted -->
<content>
<!--markup omitted -->
  <crewDrill>
    <title>Computer</title>
    <crewDrillStep>
      <challengeAndResponse>
        <challenge><para>Computer Display</para></challenge>
        <response>
          <para>
            <captionGroup cols="2" applicRefId="app-0001" changeMark="1"
             reasonForUpdateRefIds="chg-0001">
```

```
<!--markup omitted -->
            </captionGroup>
          </para>
        </response>
      </challengeAndResponse>
    </crewDrillStep>
  </crewDrill>
<!--markup omitted -->
</content>
```

Defining product attributes with the ACT

The ACT declares product attributes with the `<productAttribute>` element in the content section, and references the CCT and PCT through the `<condCrossRefTableRef>` and `<productCrossRefTableRef>` elements respectively. This is illustrated in the following example:

```
<applicCrossRefTable>
  <productAttributeList>
    <productAttribute id="serialno">
      <name>Serial number</name>
      <displayName>SN</displayName>
      <descr>Serial number etched on the bicycle frame</descr>
    </productAttribute>

<!--markup omitted -->

</productAttributeList>
<condCrossRefTableRef>
    <dmRef>
      <dmRefIdent>
        <dmCode modelIdentCode="S1000DBIKE" systemDiffCode="AAA"
         systemCode="D00" subSystemCode="0" sub-subsystemCode="0"
         assyCode="00" disassyCode="00" disassyCodeVariant="AA"
         infoCode="00Q" infoCodeVariant="A" itemLocationCode="D"/>
      </dmRefIdent>
    </dmRef>
  </condCrossRefTableRef>
  <productCrossRefTableRef>
    <dmRef>
      <dmRefIdent>
        <dmCode modelIdentCode="S1000DBIKE" systemDiffCode="AAA"
         systemCode="D00" subSystemCode="0" sub-subsystemCode="0"
         assyCode="00" disassyCode="00" disassyCodeVariant="AA"
         infoCode="00P" infoCodeVariant="A" itemLocationCode="D"/>
      </dmRefIdent>
    </dmRef>
  </productCrossRefTableRef>
</applicCrossRefTable>
```

Specifying conditions in the CCT

The CCT declares conditions that can affect applicability of data. In the content section, the <condType> element defines types of conditions, while <cond> elements define specific conditions that belong to a particular type. As shown in the following example, you can define a generic service bulletin type with allowable values of Pre and Post, then a specific service bulletin condition to determine whether, for example, a chain guard service bulletin has been installed (Post) or not (Pre):

```
<condCrossRefTable>
  <condTypeList>
    <condType id="SB">
      <name>Service bulletin</name>
      <descr>Generic service bulletin type</descr>
      <enumeration applicPropertyValues="Pre|Post"/>
    </condType>
  </condTypeList>
  <condList>
    <cond condTypeRefId="SB" id="SB-S001">
      <name>Service bulletin S001 - Chain guard</name>
      <descr>Service bulletin S001 for chain guard installation</descr>
    </cond>
  </condList>
<incorporation>
    <condIncorporation condRefId="SB-S001">
      <documentIncorporation>
        <refs>
          <dmRef>
            <dmRefIdent>
     <!--markup omitted -->
            </dmRefIdent>
          </dmRef>
        </refs>
        <incorporationStatus incorporationStatus="incorporated"
        year="2007" month="07" day="31"/>
      </documentIncorporation>
    </condIncorporation>
  </incorporation>
</condCrossRefTable>
```

The optional incorporation status list element <incorporation> is used to document technical conditions that have been incorporated into the associated publications or data modules. It can reference a condition defined in the CCT and provide a list of documents the condition affects. The list can include a reference to the data module or publication affected and the status of the incorporation, including optional date and applicability information to identify which product instances are affected.

Specifying product instances in the PCT

The PCT defines product instances and associates actual values to product attributes and conditions for each product instance. It contains a list of assignments containing the following information:

- A reference to the product attribute or condition (the `applicPropertyIdent` attribute)
- Whether this is a product attribute or condition (the `applicPropertyType` attribute)
- The actual value (the `applicPropertyValue` attribute).

The following show an example of PCT coding:

```
<productCrossRefTable>
  <product>
    <assign applicPropertyIdent="serialno"
     applicPropertyType="prodattr" applicPropertyValue="2C080754"/>
    <assign applicPropertyIdent="model"
     applicPropertyType="prodattr" applicPropertyValue="Yamazuki"/>
    <assign applicPropertyIdent="version"
     applicPropertyType="prodattr" applicPropertyValue="350"/>
  </product>
<!--markup omitted -->
</productCrossRefTable>
```

Publications and deliverable media

One definition of *publication* from the S1000D specification is:

> *'The compilation and publishing of information for a customer. This can be an IETP, a paper publication compiled from DMs or a publication containing legacy data. The List of Applicable Publications (LOAP) lists the required publications for a customer project.'*

For each customer it must be decided which types of publication are delivered, whether it consists of page-oriented publications or IETPs, and the deliverable media. The front matter for each type of publication must also be agreed. The content of the front matter depends on the publication medium and the content of the publication: for example, an access illustration data module can be included for an IETP, while a title page will be required for a page-oriented publication. The front matter can also contain, for example, a TOC and tables of symbols, terms and abbreviations if these are required. For each type of front matter, the S1000D specification gives the DMC for the data module containing the front matter.

The List of Applicable Publications (LOAP) is a publication - corresponding to an information set - that lists all the technical publications and documents for a product, a project, or part of a project. The LOAP is often used as a contractual document for the delivery of publication packages.

Publication modules

To organize the data modules for a particular publication, an XML file called a *publication module* is used. It can contain references to:

- Data modules, including front matter data modules and access illustration data modules. This can include multiple references to a single date module.
- Other publication modules.
- Technical publications that are not in S1000D format (referred to in the specification as *legacy publications*).

The publication module can have three parts:

- An optional XML RDF/disassembly code metadata section, contained in the `<rdf:Description>` element.
- A mandatory publication module identification and status section, contained in the `<identAndStatusSection>` element, which is similar to the identification and status section in a data module.
- A mandatory publication module content section (`<content>` element) containing the references to data modules, publication modules or non-S1000D publications in the order and structure in which the publication is delivered.

Within the content section of a publication module, the `<pmEntry>` element is the central element. It can contain the child elements:

- `<pmEntryTitle>`: the title of the publication entry, such as 'Front Matter', 'Chapter' and so on. The titles can be used to generate the table of contents.
- `<dmRef>`: a reference to a data module.
- `<pmRef>`: a reference to a publication module.
- `<externalPubRef>`: a reference to a non-S1000D publication or document, for example a PDF file. You can also include non-S1000D publications by encapsulating them in a data module referenced by a `<dmRef>`.

- <pmEntry>: a container for nesting other publication modules. Depending on how these elements are nested, the <pmEntry> can correspond to front matter (TOC, list of tables and so on), a chapter, section or subsection.

The following markup example represents a valid publication module fragment in XML format:

```
<pm xmlns:xsi="http://www.w3.org/2001/XMLSchema-instance"
xsi:noNamespaceSchemaLocation="http://www.s1000d.org/S1000D_4-
0/xml_schema_master/pm/pmSchema.xsd"
xmlns:rdf="http://www.w3.org/1999/02/22-rdf-syntax-ns#"
xmlns:dc="http://www.purl.org/dc/elements/1.1/"
xmlns:xlink="http://www.w3.org/1999/xlink">
  <identAndStatusSection>
    <pmAddress>
      <pmIdent>
        <pmCode pmVolume="00" modelIdentCode="DEE1B" pmNumber="01132"
         pmIssuer="C0419"/>
        <language countryIsoCode="US" languageIsoCode="sx"/>
        <issueInfo inWork="01" issueNumber="002"/>
      </pmIdent>
      <pmAddressItems>
        <issueDate day="01" month="01" year="2010"/>
        <pmTitle>Air vehicle maintenance - Landing gear system</pmTitle>
      </pmAddressItems>
    </pmAddress>
    <pmStatus>
    <!--markup omitted -->
    </pmStatus>
  </identAndStatusSection>
  <content>
   <pmEntry>
     <pmEntryTitle>Front matter</pmEntryTitle>
     <dmRef>
       <dmRefIdent>
       <!--markup omitted -->
       </dmRefIdent>
     </dmRef>
     </pmEntry>
    <pmEntry>
      <pmEntryTitle>Extension and retraction</pmEntryTitle>
      <dmRef>
        <dmRefIdent>
        </dmRefIdent>
      </dmRef>
   </pmEntry>
  </content>
</pm>
```

Publication modules are processed using software applications and style sheets that identify the data modules required for a publication and publish the required documents, whether they are page-oriented manuals or electronic publications. In an IETP environment, a navigable table of contents can be generated based on the structure of the publication module. To include non-S1000D publications in an IETP, the software must be able to read and present the information.

A SCORM (Shareable Content Object Reference Model) content package is similar to a publication module, and is used to organize learning data modules for SCORM learning products. The content section of the SCORM content package contains references to data modules, graphics and multimedia, and other SCORM content packages in the order and structure within which the learning product is delivered.

Common source database

According to the S1000D specification, the common source database (CSDB) is:

> *'An information store and management tool for all objects required to produce the technical publications within projects.'*

The information objects that the CSDB stores and manages are as follows:

- *Data modules.* The DMC and other metadata in the identification and status part of data modules, used for management purposes.
- *Illustrations and multimedia files.* A variety of file formats are supported for illustrations, including CGM4 (ATA profile), CALS Raster Group 4, JPEG, GIF, PNG, PDF and TIFF. An illustration control number (ICN) is used to identify and manage the illustrations in the CSDB. The ICN contains similar information to the identification and status section of a data module, for example, security classification and issue number.
- *Data module lists.* The DMRL and CSDB status list.
- *Comments.* Quality assurance comments associated with data modules or publication modules. The comment was known as an 'in process review form' (IPRF) in earlier issues of S1000D.
- *Publication modules and SCORM content packages.* The publication module code and SCORM module code respectively of these modules, used for management purposes.

- *Data dispatch notes (DDN).* The DDN is the standard exchange mechanism for sharing data among project members, business partners or differing CSDB systems.

All of the above information objects, apart from illustrations and multimedia files, must be produced in XML.

The S1000D specification insists on data integrity and security for all actions taken on information objects, but it does not stipulate the design and implementation rules for a CSDB. The DMC and data module metadata facilitate management and quality assurance of objects, for example by allowing subsets of information to be chosen by query, but how this is achieved is dependent on the implementation of the CSDB.

S1000D and reuse

S1000D, with its concept of data modules, is clearly well suited to reuse. Items of information that are repeated in different contexts, such as warnings or opening and closing procedures, can be stored once as a single data module and used many times in different contexts. This provides considerable savings in data maintenance and enhances data configuration control.

In S1000D, you can reuse content in a number of ways:

- By reusing data modules and publication modules within publication modules
- By reusing data from technical information repository modules
- By reusing alternate data modules referred to by a container data module.

This reuse eliminates information redundancy and produces content that is more consistent and easier to maintain, saving time and money. Furthermore, material is easier to share between products, groups and companies.

Reusing data modules and publication modules

You can reuse the same data module or set of data modules in different publication modules for different output formats, or for different purposes. You can also reuse a data module in several places within a publication module.

Apart from reusing data modules in a S1000D publication module, you can also reuse whole publication modules. This is useful if you have an existing publication module and you want to extend it for a particular purpose, or, for example, when you want to use the same set of data modules in different publications.

Reuse through the technical information repository data module

A project or organization can decide to store detailed information in technical information repository data modules. Examples of such data include functional items, circuit breakers, parts, access points, supplies, support equipment and controls and indicators. This allows detailed technical information to be stored in a controlled manner and shared and reused easily in data modules across a whole project. There can be a technical information repository module for each type of data.

As an example, a data module might contain a reference to a functional item as follows:

```
<functionaltemRef
   functionaltemNumber="202B12" manufacturerCodeValue="GAPB6">
</functionaltemRef>
```

The markup in the technical information repository might be as follows:

```
<content>
  <techRepository>
    <functionalItemRepository>
   <!--markup omitted -->
    <functionalItemSpec id="202B12 GAPB6">
      <funcionalItemIdent
       functionaltemNumber="202B12" manufacturerCodeValue="GAPB6"/>
      <name>CTL MODULE-VENT FWD</name>
    </functionalItemSpec>
    </functionalItemRepository>
</techRepository>
</content>
```

During the publication process, detailed data such as the functional item name – 'CTL MODULE-VENT FWD' in this example – is retrieved from the technical information repository and included in each referencing data module. This is an example of an implicit reference, but explicit references, in which the corresponding technical information repository data module is explicitly referenced using the `<dmRef>` element, can also be used.

Reuse through the container data module

A container data module associates alternate data modules such that references to the container data module target the group of alternate data modules. You use a container data module, for example, when several data modules achieve the same maintenance goal, but the detailed procedures differ due to product configuration, maintenance environment or other conditions. Processing software that generates IETPs or other publications should hide the container data module, while displaying the appropriate alternate data module on output.

A container data module limits the impact of configuration changes. Without a container data module, you would need to create a new issue of any referencing data module to reference both of the alternate data modules (that is, the `<dmRef>` would need to change). With a container data module a new issue of the referencing data module is not necessary, because the reference to the container data module remains valid (see Figure 4.3).

A container data module cannot refer to another container data module, and can only refer to data modules with content.

As shown in the following example, the container data module and the alternate data modules are usually distinguished by the disassembly code variant with values A, B and C:

```
<dmodule>
  <identAndStatusSection>
    <dmAddress>
      <dmIdent>
        <dmCode modelIdentCode="AJ" systemDiffCode="A"
         systemCode="35" subSystemCode="1" subSubSystemCode="3"
         assyCode="51" disassyCode="00" disassyCodeVariant="A"
         infoCode="720" infoCodeVariant="A" itemLocationCode="A"/>
<!--markup omitted -->
      </dmIdent>
<!--markup omitted -->
    </dmAddress>
<!--markup omitted -->
      <applic><displayText>All</displayText></applic>
<!--markup omitted -->
  </identAndStatusSection>
  <content>
    <container>
      <refs>
       <dmRef>
         <dmRefIdent>
           <dmCode modelIdentCode="AJ"
            systemDiffCode="A" systemCode="35"
            subSystemCode="1" subSubSystemCode="3" assyCode="51"
            disassyCode="00" disassyCodeVariant="B" infoCode="720"
            infoCodeVariant="A" itemLocationCode="A"/>
         </dmRefIdent>
       </dmRef>
```

```
      <dmRef>
        <dmRefIdent>
          <dmCode modelIdentCode="AJ"
           systemDiffCode="A" systemCode="35"
           subSystemCode="1" subSubSystemCode="3" assyCode="51"
           disassyCode="00" disassyCodeVariant="C" infoCode="720"
           infoCodeVariant="A" itemLocationCode="A"/>
        </dmRefIdent>
      </dmRef>
     </refs>
   </container>
  </content>
</dmodule>
```

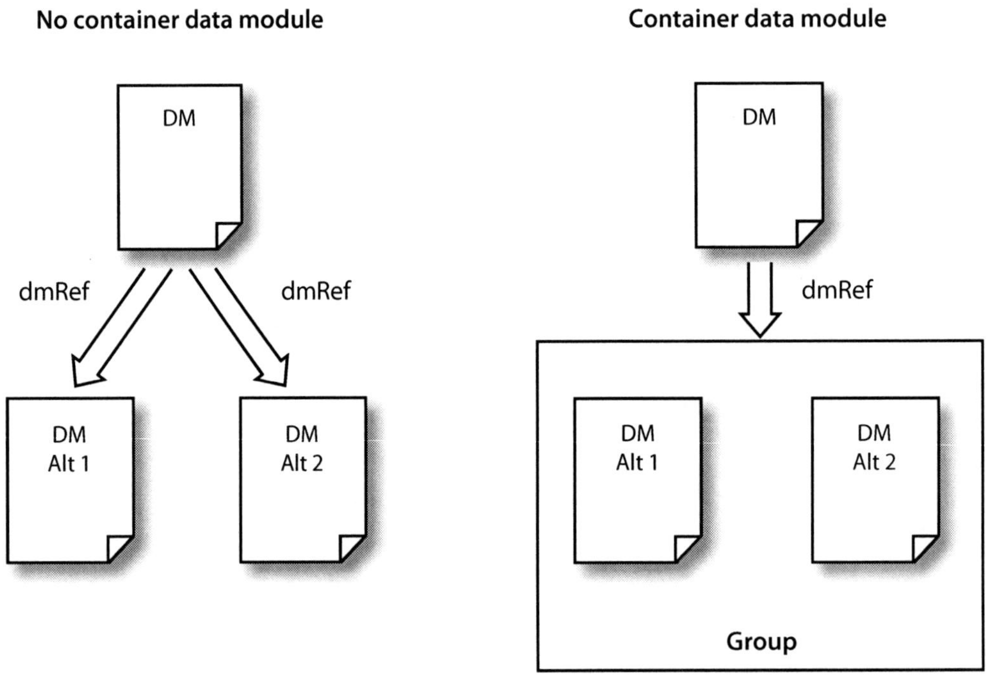

Figure 4.3 Container data module

Tools available for working with S1000D

You can use an XML editor to produce S1000D code, ideally using an S1000D-specific plug-in to ensure that code conforms to the specification and business rules used for the project. For example, the Eclipse S1000D for PTC Arbortext Editor product from Mekon allows you to produce S1000D compliant output; Mekon offer other products for

S1000D authoring, such as Eclipse Create S1000D for FrameMaker. PTC has its own Arbortext Editor plug-in, the Authoring Tool Interface.

In FrameMaker, support for S1000D is available from Version 7.2 by means of an application pack, but this only includes applications for the following data modules:

- Business rules exchange
- Description
- Illustrated parts data
- Procedural.

As the FrameMaker application pack for S1000D also includes a set of schema documents for other S1000D data modules, it is feasible to extend the application pack to create a new S1000D application. The application pack allows you to import and export S1000D XML source files using a structured application, with the associated template, read/write rules file and other files. As of 2010 the application pack is still a beta release and only supports S1000D schemas up to Issue 2.2.1.

There are a number of products that not only support S1000D authoring, but also offer CSDB implementations, thus helping manage the documentation process. For example:

- S1000D CSDB for Sibersafe provides facilities for creating, managing and delivering S1000D-compliant documentation.
- Inmedius S1000D Publishing Suite is another end-to-end software solution.
- Arbortext Common Source Database (CSDB) software for S1000D is a complete CMS that supports the S1000D specification. Together with Arbortext Learning Content Manager for S1000D, Arbortext LSA Interface for S1000D and Arbortext Parts Catalog Manager for S1000D, it forms a suite of S1000D products from PTC.

These solutions allow creation and management of business rules and data modules and publishing of S1000D technical publications for both print and IETP viewing.

The level of support for S1000D varies between products, and the latest issue of the S1000D specification may not be supported. The S1000D information resource site `www.s1000d.net` maintains a list of software suppliers.

Advantages of S1000D

Due to the role it gives data modules, S1000D is well-suited to single-source publishing. Information can be created once and reused in many publications and projects, and content is easily repurposed for publication in both paper-based and electronic formats. It also offers other facilities that allow reuse of information: the technical information repository data module and the container data module. All this brings obvious advantages of time and cost-saving and maintainability, reduced translation cost and the added advantages of increased consistency and accuracy.

S1000D is accepted internationally across the aerospace and defence industries. It is non-proprietary, so you are not tied into the products of a single vendor. This neutrality and the platform-independence afforded by XML allows easier information exchange between organizations in the supply chain and throughout the lifecycle of the product.

The concept of the CSDB and data module code allows for easier management of content and project materials. Information is filtered easily, as metadata aids in searching and retrieving data. Legacy data can be migrated to the CSDB and either encapsulated in S1000D data modules or referenced from publication modules.

S1000D provides schema to support a large number of data module types, yet through the use of business rules it can be configured to meet specific customer requirements for a project. In addition, the use of applicability allows the published documentation to be tailored for different configurations.

S1000D and DITA compared

If you are involved in military and aerospace projects, you are very likely to use S1000D – indeed, its use may be mandated: for software documentation projects you are more likely to use DITA. There may, however, be documentation projects in which you have the freedom to choose either solution, so it is worthwhile discussing the similarities and differences between the two standards.

S1000D and DITA have a number of similarities:

Both have self-contained reusable units of information, data modules in S1000D, topics in DITA, and therefore both support topic-based reuse and repurposing of content. Both also provide other reuse mechanisms: in DITA, reuse of publication modules/DITA maps; in S1000D, the reuse facilities offered by the container data module and technical information repository data module.

- In both S1000D and DITA, content is categorized by specific information types. S1000D has the more generic description and procedural data module types and a range of data module types specifically for maintenance, fault isolation, crew operations and so on. On the other hand, DITA has the Concept, Task and Reference topic types, while the generic Topic type can be specialized as required.
- Both provide mechanisms for organizing content for output. In S1000D, a publication module is used to list data modules for building publications, while DITA uses DITA maps, which can structure content hierarchically or by relationship to other topics.
- Both support the addition of metadata about data modules or topics, S1000D with the status and identification section and DITA with its metadata attributes and prolog section in topics.
- Both allow conditional processing, S1000D through its applicability feature and DITA through its filtering and flagging mechanisms.

DITA has a number of features that S1000D does not have:

- *Specialization.* S1000D has a set number of data module types, but with DITA you can exploit structural specialization to create new information types as specializations of existing types. With domain specialization, you can also add new elements in domain-specific vocabularies. In contrast, the elements in S1000D are predefined, though you can decide through business rules which ones to use.
- *Content referencing.* The `conref` attribute of DITA allows very granular reuse of information. Admittedly, later issues of S1000D do allow granular reuse through the technical information repository mechanism, and some S1000D editors, such as Arbortext CSDB for S1000D, allow very granular reuse of information.

S1000D has a number of features that DITA does not have:

- *The CSDB.* The S1000D specification includes this mechanism for the storage and management of documentation content and deliverables. The DMC and other metadata allow data to be easily identified and managed in the CSDB.

- *Business rules.* In the business rules for a project you can determine, for example, which elements are used, as well as their interpretation, to what extent applicability is used and so on.
- *The process data module.* This data module type can present interactive procedural flows, including branching, looping and filtering, across a set of related content modules, and can allow interaction with external software.

In summary, you cannot add to the set of elements available in S1000D, although projects can customize the way they use the specification by way of business rules. This promotes interoperability and means that content is more easily maintained. Business rules exchanged between organizations result in a common understanding of how the specification is interpreted for a project. On the other hand, DITA provides a core set of elements and topic types, but is designed to be customizable. Through specialization, you can customize it for your specific requirements. In S1000D you can choose the elements that you use, whereas with DITA you can create new elements according to your requirements.

S1000D data module types mainly support the industries in which S1000D is used, hence the data module types for aircraft maintenance procedures and military equipment instructions. Using specialization you could potentially add information types in DITA corresponding to the S1000D information types, but this would be a large undertaking.

Both S1000D and DITA are widely supported; DITA may offer a simpler starting point, while S1000D may be more expensive to implement. While there are projects where you might have the freedom to choose between S1000D and DITA, you are very likely to use S1000D for military and aerospace projects, but for software documentation projects, DITA is the more likely choice.

Figure 4.4 Abridged version of sample data module DMC-S1000DBIKE-AAA-DA0-20-00-00AA-520A-A_006-00_EN-US.xml

```
<?xml version="1.0" encoding="UTF-8" ?>
<!DOCTYPE dmodule
[
]>
  <dmodule xsi:noNamespaceSchemaLocation="http://www.s1000d.org/S1000D_4-
0/xml_schema_flat/proced.xsd"
xmlns:dc="http://www.purl.org/dc/elements/1.1/"
xmlns:rdf="http://www.w3.org/1999/02/22-rdf-syntax-ns#"
xmlns:xlink="http://www.w3.org/1999/xlink"
xmlns:xsi="http://www.w3.org/2001/XMLSchema-instance">
 <rdf:Description>

 <!--markup omitted -->

 </rdf:Description>
  <identAndStatusSection>
  <!-- Identification part -->
  <dmAddress>
    <dmIdent>
      <dmCode modelIdentCode="S1000DBIKE" systemDiffCode="AAA"
       systemCode="DA0" subSystemCode="2" subSubSystemCode="0"
       assyCode="00" disassyCode="00" disassyCodeVariant="AA"
       infoCode="520" infoCodeVariant="A" itemLocationCode="A">
      </dmCode>
      <language countryIsoCode="US" languageIsoCode="en"/>
      <issueInfo issueNumber="006" inWork="00"/>
    </dmIdent>
    <dmAddressItems>
      <issueDate year="2008" month="08" day="01"/>
      <dmTitle>
        <techName>Rear wheel</techName>
        <infoName>Remove procedures</infoName>
      </dmTitle>
    </dmAddressItems>
  </dmAddress>
  <!-- Status part -->
  <dmStatus issueType="revised">
    <security securityClassification="01"
     commercialClassification="cc51" caveat="cv51" />
   <!--markup omitted -->
    <responsiblePartnerCompany enterpriseCode="U8025">
      <enterpriseName>UK MoD</enterpriseName>
    </responsiblePartnerCompany>
    <originator enterpriseCode="U8025">
      <enterpriseName>UK MoD</enterpriseName>
    </originator>
    <applicCrossRefTableRef>
```

Figure 4.4 Abridged version of sample data module DMC-S1000DBIKE-AAA-DA0-20-00-00AA-520A-A_006-00_EN-US.xml (continued)

```
<dmRef xlink:type="simple" xlink:actuate="onRequest"
      xlink:show="replace"
      xlink:href="URN:S1000D:DMC-S1000DBIKE-AAA-D00-00-00-00AA-00WA-D">
       <dmRefIdent>
        <dmCode modelIdentCode="S1000DBIKE" systemDiffCode="AAA"
         systemCode="D00" subSystemCode="0"
         subSubSystemCode="0" assyCode="00" disassyCode="00"
       disassyCodeVariant="AA" infoCode="00W"
       infoCodeVariant="A" itemLocationCode="D" />
     </dmRefIdent>
    </dmRef>
   </applicCrossRefTableRef>
   <applic>
     <displayText>
       <simplePara>Mountain bicycle and (Mountain storm Mk1 or
        Brook trekker Mk9)</simplePara>
     </displayText>
     <evaluate andOr="and">
       <assert applicPropertyIdent="type"
        applicPropertyType="prodattr"
        applicPropertyValues="Mountain bicycle"/>
     <evaluate andOr="or">
     <evaluate andOr="and">
     <assert applicPropertyIdent="model"
      applicPropertyType="prodattr"
      applicPropertyValues="Mountain storm" />
     <assert applicPropertyIdent="version"
      applicPropertyType="prodattr" applicPropertyValues="Mk1" />
     </evaluate>
     <evaluate andOr="and">
     <assert applicPropertyIdent="model"
      applicPropertyType="prodattr"
      applicPropertyValues="Brook trekker" />
     <assert applicPropertyIdent="version"
      applicPropertyType="prodattr" applicPropertyValues="Mk9" />
     </evaluate>
     </evaluate>
     </evaluate>
   </applic>
  <!--markup omitted -->
   <brexDmRef>
     <dmRef xlink:type="simple" xlink:actuate="onRequest"
      xlink:show="replace"
```

Figure 4.4 Abridged version of sample data module DMC-S1000DBIKE-AAA-DA0-20-00-00AA-520A-A_006-00_EN-US.xml (continued)

```
xlink:href="URN:S1000D:DMC-S1000DBIKE-AAA-D00-00-00-00AA-022A-
      D_007-00">
       <dmRefIdent>
        <dmCode modelIdentCode="S1000DBIKE" systemDiffCode="AAA"
         systemCode="D00" subSystemCode="0"
         subSubSystemCode="0" assyCode="00" disassyCode="00"
         disassyCodeVariant="AA" infoCode="022"
         infoCodeVariant="A" itemLocationCode="D" />
        <issueInfo issueNumber="007" inWork="00" />
      </dmRefIdent>
     </dmRef>
   </brexDmRef>
   <qualityAssurance>
     <firstVerification verificationType="tabtop" />
   </qualityAssurance>
   <!--markup omitted -->
   <reasonForUpdate>
     <simplePara>totally revised</simplePara>
     <simplePara>Schema cleanup element/attribute renaming</simplePara>
   </reasonForUpdate>
  </dmStatus>
</identAndStatusSection>
  <content>
   <procedure>
     <preliminaryRqmts>
       <reqCondGroup> <noConds /> </reqCondGroup>
       <reqPersons><personnel /></reqPersons>
       <reqSupportEquips><noSupportEquips /></reqSupportEquips>
       <reqSupplies><noSupplies /></reqSupplies>
      <reqSpares><noSpares /></reqSpares>
      <reqSafety><noSafety /></reqSafety>
    </preliminaryRqmts>
    <mainProcedure>
      <proceduralStep>
        <para>Hold the rear of the bicycle.</para>
      </proceduralStep>
      <proceduralStep>
        <para>Push the wheel forwards and down to disengage
         the chain from the sprocket.</para>
      </proceduralStep>
      <proceduralStep>
        <para>Turn the wheel to the side and lift it
         away from the frame.</para>
      </proceduralStep>
      <proceduralStep>
        <para>Put the frame on the floor.</para>
      </proceduralStep>
```

Figure 4.4 Abridged version of sample data module DMC-S1000DBIKE-AAA-DA0-20-00-00AA-520A-A_006-00_EN-US.xml (continued)

```
  </mainProcedure>
      <closeRqmts>
        <reqCondGroup><noConds /></reqCondGroup>
      </closeRqmts>
    </procedure>
  </content>
</dmodule>
```

5

Authoring with XML

This chapter discusses the tools that you can use to create and edit XML documents:

- Simple text editors.
- XML editors: tools dedicated to XML authoring that use XML as a native format.
- XML-aware tools: tools such as FrameMaker that provide structured authoring and XML output from a 'traditional' authoring interface.
- Content management systems (CMS): software that provides content management functionality and which include XML authoring and publishing capabilities.

Text editors

XML is text, therefore you can use simple text editors such as Notepad to work with XML files. Such editors are free and easy to use, but do not provide any facilities for rapid entry, for validation of markup or for showing structural views of an XML document.

Unless you are particularly adept at typing markup, you will prefer an editor that does much of the work for you. As such, simple text editors may only be suitable for programmers or those used to working directly with markup languages.

XML editors

There are many XML editors available on the market and their cost and functionality vary considerably. There are fairly simple free editors that you can download for yourself, through to expensive full-function packages that provide the facilities that you would expect from high-end authoring tools.

At the lower end you can download Microsoft's XML Notepad free from `www.microsoft.com`. XML Notepad is a simple but effective editor that provides some assistance in entering markup and validates documents as you type. It divides the screen into two panes: in the left pane, it shows a structural view of the elements in the document, while showing the content of each element in the right-hand pane. It also provides an XSL output view to show what the XML documents look like after transformation with an XSLT stylesheet. XML Notepad is probably only suitable for people who want to dabble in XML, rather than those who want to use XML for producing documentation.

Most authors are likely to require a more sophisticated authoring environment, an editor that can use stylesheets to present the text content and hide the intricacies of the markup if required. You might also expect a spell-checker, table tools, book building tools and other functions to support your document authoring. There are a number of tools that provide such functionality, including: XMetal, XMLSpy, PTC Arbortext Editor (formerly Epic) and MadCap Flare.

There are certain features that you will want in an XML editor, so when you are evaluating an editor you should check that these features are provided. If the editors listed above do not provide all of these features, it is likely that they will do so in the near future.)

A checklist of desirable features of XML editors is shown in Table 5.1.

Table 5.1 Desirable features in XML editors

Feature	*Description and considerations*
Formatting views	Multiple formatting views: ■ A WYSIWYG view or visual editor ■ A structure tree view showing the element hierarchy with coloured highlighting of element names, attributes and entities ■ An XML code view. The view you prefer is likely to be a matter of experience and personal preference. Some users will want to author documents without requiring any great knowledge of the XML language or XML programming.

Table 5.1 Desirable features in XML editors (continued)

Feature	***Description and considerations***
Assisted entry of markup	Easy entry of markup, for example by: ■ Menus and toolbars, allowing users to choose elements and attributes from a list or drop-down menu ■ Wizards ■ Templates for, say, table structures ■ Drag and drop of structures such as an element and its children.
Validity checking	Checking of well-formedness and validity and enforcement of correct structure as it is entered. The editor should not allow the user to make tagging errors and should show which elements and attributes are valid at the current cursor position.
Support for DTD and schemas	Support for standard XML schema, including the XML language that you intend to use. Are DocBook, DITA and S1000D supported? Can you create or customize DTDs or schemas? Is output validated against the DTD or schema?
Importing of text and tables	Importing or cutting and pasting of material from other popular formats such as Microsoft Word or CSV files. Can you import tables from Word? Are there wizards for assisting with importing?
Support for XSL	Ability to transform XML to HTML using stylesheets and view the output. Ability to create and edit XSLT stylesheets. Are XSLT, XSL-FO and XPath supported and are all the features of these standards supported?
Support for natural languages	Support for authoring in languages other than English, including double-byte character set (DBCS) languages.
Content management	Provision of content management facilities such as file storage and version control, or integration with a separate content management system.

Table 5.1 Desirable features in XML editors (continued)

Feature	*Description and considerations*
Publishing of output formats	Publishing of XML source documents to output formats such as HTML and PDF, or integration with separate publishing tools. Preview of output.

The website `XML.com` has a considerable list of XML editors, complete with product information, reviews and links to vendor websites.

Apart from the XML-related features listed in the table, you will also want standard authoring features such as spell-checking, global search and replace, table tools, revision markers, macros and all the usual functions of a text editor. You might however not require all of the features in Table 5.1, depending on your intended usage of XML. For example, while the assisted entry of markup and validity checking features are probably the most important, you might not need to create or edit XSLT stylesheets or DTDs. Of course, the cost of the XML editor is a major consideration, as is the time and effort that might be involved in customizing the tool and associated DTDs for your use and providing training on the new tools.

Some XML editors, for example XMLSpy, are more suitable for programmers than for technical communicators, as they provide features more useful for software development rather than for building documentation.

Another XML editor, and one that takes a rather different approach to editing, is Altova's Authentic Editor. It is an XML and database content editor that allows you to edit documents using a word processor-style interface that presents the documents as forms. You can edit database content directly without being exposed to the underlying technology, although you do probably need an understanding of relational databases to use this feature of the editor. Authentic is available free of charge and comes in Desktop and Browser editions. The Browser edition is a plug-in that runs within Internet Explorer 5.5 and higher.

XML and FrameMaker

In FrameMaker, you can work in a structured or unstructured authoring environment. Before FrameMaker 7, the structured version of FrameMaker was sold as FrameMaker+SGML, while the unstructured version was the regular FrameMaker product. From Version 7, however, the two environments are delivered in the same product.

In FrameMaker, a structured document is one that uses a defined structure based on an Element Definition Document (EDD), which is the equivalent of a DTD in FrameMaker. An unstructured document is a FrameMaker document that does not use an EDD-based structure, and which uses paragraph and character tags to handle formatting. In a structured document, formatting instructions can be included in the EDD itself, or the EDD can refer to a document's paragraph and character tags to handle formatting.

Although FrameMaker does not support XML as a native format, it does allow you to:

- Export XML from unstructured FrameMaker documents.
- Import and export XML to and from structured FrameMaker documents. When you import, the XML elements are represented in the element hierarchy of the structured FrameMaker document. Importing and exporting XML is known as *round tripping.*

You export files to XML using the **File → Save As** command, and you import XML files using the **File → Open** command. When you import a set of XML files in which a master file embeds other files using entities, you can create a whole FrameMaker book with separate documents for each of the XML files. Apart from importing XML files to create separate documents in FrameMaker, you can import XML documents as text insets in FrameMaker files using the **File → Import** option.

When you import XML files you can edit the files in FrameMaker, or simply use FrameMaker as a print engine to produce PDF or printed books. Whether or not you use XML as your source format, you can take advantage of FrameMaker's multi-channel publishing capabilities to produce PDF, HTML and online help formats. In particular, FrameMaker can offer high quality PostScript and hence PDF output. See Figure 5.1.

You might want to export to XML because you want to use XML as a storage format in your content repository. There is the added advantage of being able to add metadata to the files that you store. XML files, because they are text files, are easier to share than large binary files in FrameMaker's native binary format. Alternatively, rather than export XML from your FrameMaker files, you can also create XML using the version of WebWorks Publisher that is packaged with FrameMaker.

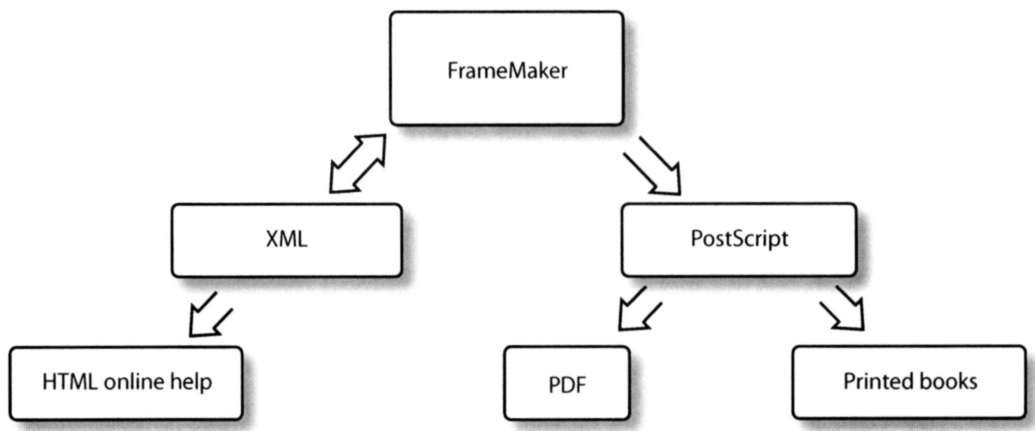

Figure 5.1 FrameMaker XML import and export

Exporting XML from unstructured FrameMaker

To export to XML from unstructured FrameMaker, you save the files as XML. When you save a book, one or more XML output files are created for each FrameMaker file, together with a corresponding Cascading Style Sheet (CSS) that can be used with the document.

When you export XML in this way, the conversion settings for the book are stored on the reference pages in the first file in the book. These settings define the mappings between FrameMaker paragraph and character tags and XML elements. To modify the conversion settings you must edit the reference pages.

The exported XML has element names based on your FrameMaker tags, which may not be what you require, but it is always possible to transform the XML to any XML vocabulary that might be required.

Using XML with structured FrameMaker

To import and export XML to and from structured FrameMaker documents, it is best to use a *structured application* (also called a *structure application*), which is a list specifying the location of the files that control how your FrameMaker documents are transformed into XML and vice versa.

The structured application file supplied with FrameMaker (`structapp.fm`) includes definitions for several structured applications, including those for DocBook, DITA and XHTML. The structured applications available depend on the version of FrameMaker in use: for example, a DocBook XML application has been available since FrameMaker 7.0, and DITA and S1000D support from FrameMaker 7.2.

You can use the supplied structured applications as-is or you can customize them. If you customize, it is advisable to first copy the definitions and then update settings as required. The other possibility is to create your own structured application after preparing your own DTD, template file and other related files.

Creating and editing structured applications is not particularly straightforward, as it requires a good knowledge of FrameMaker and XML. For detailed information about creating structured applications, see FrameMaker documentation such as the *Structure Application Developer Guide*. (The exact title of this document varies with the version of FrameMaker).

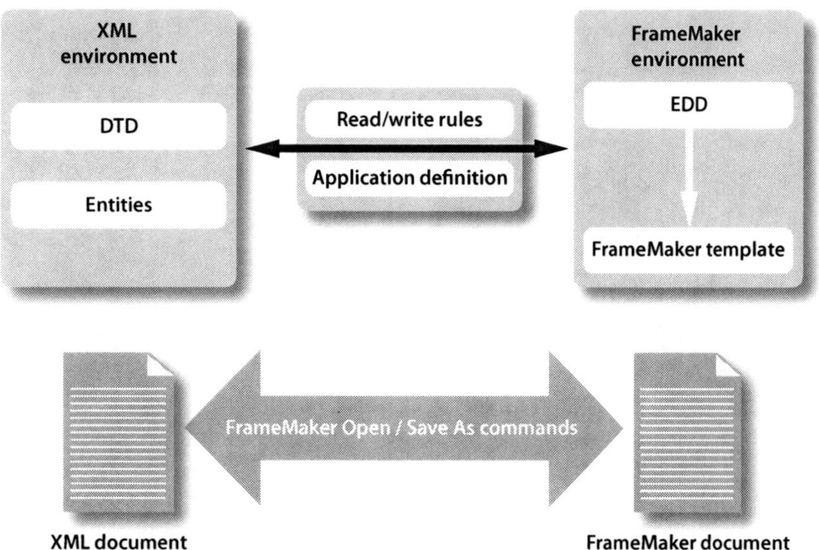

Figure 5.2 Structured applications

A structured application acts as the bridge between FrameMaker EDDs and XML DTDs. As shown in the example for a DocBook structured application in Table 5.2, it specifies the location of:

- The DTD for the XML language
- A FrameMaker template file

- An optional read/write rules file, for describing how elements are translated during round-tripping
- An optional CSS or XSLT stylesheet to be used when exporting XML (FrameMaker 7.2 and later)
- Entity and module files associated with the DTD (not shown in the DocBook example).

The structured application also specifies the document types that can appear at the top of XML files in the XML language and any application programming client (API) client to be used. An API client is specially written software for customizing XML output.

Table 5.2 A DocBook structured application

Application name:XDocBook
  DTD: $STRUCTDIRxmlxdocbookappdocbookx.dtd
  Template: $STRUCTDIRxmlxdocbookapptemplate
  Read/write rules: $STRUCTDIRxmlxdocbookapprules
  CSS2 Preferences:
    Generate CSS2: Disable
    Add Fm CSS Attribute To XML: Disable
  Retain Stylesheet Information: Disable
    XML Stylesheet
    Type: css
    URI: /$STRUCTDIR/xml/xdocbook/app/xdocbook.css
  Use API client: DocBook
  Namespace: Enable
  DOCTYPE: Appendix
    Article
    Bibliography

Note that FrameMaker only supports XML Schema documents in FrameMaker Version 7.0 and later.

The formatting template is necessary because the DTD contains no formatting information. The FrameMaker EDD is similar to a DTD in that it contains a definition of elements and the structure of the document, but unlike a DTD, it can contain structure-dependent rules that specify formatting for each element. With a FrameMaker structured application, the EDD is normally imported into the template so that the structure rules and formatting information coexist in the same template. When you import XML into FrameMaker, the template and its embedded EDD are used to apply formatting to the content.

You might be lucky and find that specifying a DTD and template is enough for working with XML, or you might find that the default settings do not work well enough. In this case, you can specify a read/write rules file to refine the translation of elements during round-tripping. You might want to use the rules file to:

- Identify an element in the markup as a specific type of FrameMaker element
- Adjust an element name
- Drop an element or attribute not required in FrameMaker
- Map XML entities to FrameMaker element types
- Ensure that images are treated correctly on round-tripping.

If the read/write rules features do not give you sufficient control, you may need to use XSL transformations to process your XML files during import or export, or even use custom code in an API client.

There are some aspects of FrameMaker to XML import and export that are problematic and are likely to require special attention:

- FrameMaker variables can be converted either to plain text or to the desired entities on exporting to XML, depending on your setup.
- Conditions in FrameMaker might result in relevant processing instructions being added to XML files rather than producing the desired filtering of XML content.
- Comments in XML files might be lost on import.
- Cross-references in FrameMaker might not be handled properly on export.

All of this underlines the fact that setting up FrameMaker to work with XML can demand considerable work and expertise. However, later versions of FrameMaker, in particular Version 8.0, promise to automate much of the round-tripping conversion. For example, in FrameMaker 8.0, entities are created automatically from variables on export to XML, and markers can be added for comments on import of XML.

XML and Word

In Microsoft Word 2003 and later versions you can save files to Word's own XML format. Additionally, in Microsoft Office Professional and the stand-alone version of Word, there is a basic XML editor and facilities for importing and exporting documents in other XML languages.

In Word 2003 you can save to the WordProcessingML (WordML for short) XML language. WordXML has elements based on Word styles, but is pretty unreadable for humans, as the following snippet shows:

```
... <w:pStyle w:val="Heading1"/></w:pPr><w:r><w:t>XML and Word</w:t></w:r></
w:p><w:p wsp:rsidR="0016147A" wsp:rsidRDefault="0016147A"
wsp:rsidP="0016147A"><w:pPr><w:autoSpaceDE w:val="off"/><w:autoSpaceDN
w:val="off"/><w:adjustRightInd w:val="off"/><w:rPr><w:rFonts w:cs="Arial"/
><w:lang w:val="EN-GB"/></w:rPr></w:pPr><w:r><w:t>In all versions of
Microsoft Word 2003 and later you can save files to Word's own XML language,
but only </w:t></w:r>...
```

In Word 2007, the native file format is Office OpenXML and the default file type for saving files is as a `.docx` file, which is a collection of XML files assembled together as a Zip file. One of the XML files is named `document.xml` and this holds the document content in WordML format.

Word and XML Schema Language

In Office Professional and stand-alone Word 2003 and later there is support for XML Schema, so you can import XML instance documents for those XML schemas that Word recognizes. You have to add any schemas that you want to use to the Word Schema Library, which you access through the 'Templates and Add-Ins' dialog. You can then validate instance documents against those schemas when you export them. There are limitations to the schemas you can use, however: any XML document that you want to edit in Word must have a namespace, which precludes, for example, DocBook and DITA. Neither can you use DTDs directly: you would have to convert any DTD to XML Schema language first. One example of a tool for doing this conversion is Trang, which is available from `www.thaiopensource.com/relaxng/trang.html`.

In the XML editor you can choose the data view for the document, as shown in the left-hand pane in Figure 5.3. A data view uses an XSLT stylesheet to render the document. You can have a data view for any stylesheets associated with the document in addition to the data-only data view, which corresponds to Word's default XSLT stylesheet. Once you edit the document, however, you are locked into the data view you choose initially. The XML editor in Word is fairly robust, but lacks many of the features that you might expect in a dedicated XML editor: for example, it does not validate as you type.

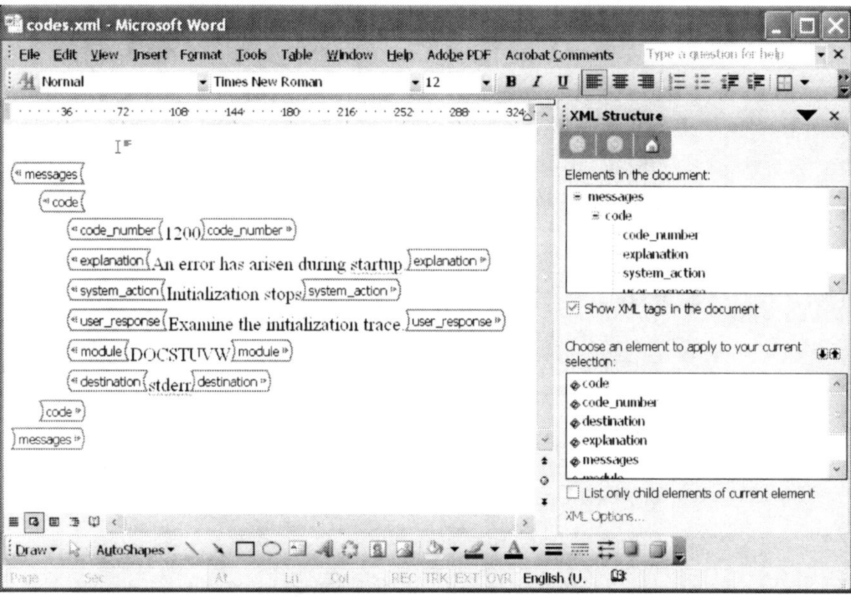

Figure 5.3 The XML editor in Word

When you import an XML document not already in WordML format, it is transformed into a WordML document, containing WordML markup merged with the markup of the original document. After editing the XML document you can save it in the WordML format to preserve the styles and formatting information, or you can select the **Save data only** option. If you select this option, the WordML tagging is removed from the document, so only the markup for the Schema language is retained. On saving you can also apply an XSLT stylesheet to the XML document to transform the document as required.

With considerable programming expertise it is possible to create a *smart document* to overcome some of the limitations of the basic XML support in Word. Smart documents have an XML expansion pack attached, which can add functionality to a Word document by specifying custom displays or actions. This would allow you to have different views of the XML document while editing, rather than just the one view you choose when opening the file, and to, for example, add actions for adding hyperlinks or boilerplate items.

In summary, XML support in Word is more about being able to share files with other software (particularly the MS Office Suite) and interfacing with databases than authoring documentation in XML format. The basic editor is limited in functionality, and there is no real support for the XML languages that you would probably want to use. Admittedly, another possibility for using Word to work with XML is to use a plug-in product, for example Quark XML Author for Word, which allows you to edit and create XML documents. Also, you could use XSLT to convert from a more useful XML vocabulary into WordML for importing, and, after exporting WordML, for converting to your required XML vocabulary.

XML and content management systems

There are various content management systems (CMS) on the market that allow you to share, manage and store XML files. Such systems typically interface with a database management system (DBMS) for storage and use a version control system for keeping track of changes to documents. XML and its role in content management is discussed in Chapter 11, *XML and Content Management.*

Some XML editors and other authoring tools provide interfaces to content management systems, allowing you to check files in and out of a CMS repository for version control. There are also CMS solutions that provide all-in-one XML authoring, content management and publishing capabilities. For example, PTC Arbortext provides a content management solution that includes an XML editor and facilities for publishing XML documents to the web, PDF and other media. Such systems as this have the necessary stylesheets, transforms, rendering engines and so on to produce automated builds from your XML content, although they can be very costly.

There are also separately-available tools that focus on XML conversion or on publishing to output formats: tools such as RoboHelp, WebWorks and DoctoHelp all provide support for XML, as do high-end composition and pagination software packages such as Arbortext Advanced Print Publisher (formerly 3B2). Whether you use separate or combined authoring, content management and publishing tools is likely to depend on the software that is already in use in your organization, as well as your budget.

6

Migrating to XML

This chapter discusses what is involved in migrating to the use of XML as the source format for your documentation. It discusses general migration considerations, as well as specific considerations for migrating from:

- HTML
- SGML
- Structured and unstructured FrameMaker
- Word.

The chapter also discusses what might be involved in migrating between different XML environments.

What is involved in migration?

Migration to the use of XML from other formats involves a lot of thought and effort. Regardless of which authoring environment you migrate from, all of the following will cost time and money:

- Choosing an XML language. Do you need the topic-oriented features of DITA, or would DocBook be a better choice? Can you use the chosen language 'out of the box' or do you need to customize it?
- Implementing an XML authoring system and publishing tools, and possibly a content management tool. Given the number of solutions on the market, choosing the most suitable solution may be time-consuming.
- Developing and maintaining DTDs or schema documents, or converting DTDs.

- Training in XML concepts and in the new tools. There will be a rather steep learning curve for those not familiar with markup languages or topic-based authoring.
- Modelling your information for the move to a structured writing environment.
- Testing any tools that you intend to use for conversion. It is advisable to run a pilot conversion project initially.
- Converting from legacy formats and validating the converted XML: there is always a lot of clean-up activity involved, even after using tools that automate much of the conversion.

Managers will need to decide whether it is more cost-effective to do the conversion in-house or to outsource the work. Outsourcing conversion work may be a more expensive option, but there are advantages to using people who are accustomed to such work and have already seen the problems of migration. It is obviously important to check the experience and track record of such organizations.

If you do the conversion in house, you will need the right people and skills: it will take up much of their time, but they will benefit from the hands-on training. Management may also have to overcome a certain amount of resistance to using new technologies or adopting new ways of working: for some, using markup language is more like programming than writing.

If the conversion work is done in-house, there is the question of how much work should be done manually as opposed to using existing or commercially available tools. And, if the tools for a particular migration are not available, there is the question of whether to wait until they are available. One approach is to cut and paste legacy content into an XML editor. This approach would allow you to do clean-up work as you go along and to learn about the markup language as you do the conversion. However, it is very time-consuming and is not really recommended.

You should consider how much of your legacy content to migrate. There is no point in migrating content that is never going to be updated or reused. Also, there is no value in migrating content to XML if it can be published more easily using the existing approach, or if the delivery format is not one that is readily supported by the XML environment to which you intend to migrate. If there are budget constraints, it makes sense to migrate only your most frequently used documentation.

The migration might involve a shift from unstructured to structured writing, for example to the topic-based authoring environment of DITA. In this case, you could apply structure before conversion by reorganizing your legacy content into appropriate chunks corresponding to topic types: content that is consistently structured is easier to convert automatically. A migration to DocBook from an unstructured legacy format would usually not involve the same reorganization as would migration to a topic-based architecture such as DITA.

Another advantage of cleaning up your documentation before conversion is that you can identify redundant and incorrect information. There is no point in paying to migrate information that is deleted soon after migration.

Of course there is a consideration of whether you really need to migrate to XML as a source format. You could retain your current source format and simply publish to XML as an output format. This may be an option to consider if you are already implementing single-sourcing.

Migrating from HTML

To convert HTML into XML, you can first convert HTML files into XHTML. You then have well-formed XML documents that you can transform into your required XML vocabulary.

You can use the HTML Tidy tool, available from `http://tidy.sourceforge.net`, to perform most if not all of the conversion to XHTML. Using a tool rather than performing the conversion manually is advisable considering the ugly formatting and hard-to-read nature of much existing HTML.

The following is a list of changes that must be made to convert HTML to well-formed, valid XHTML:

- Encode the document in UTF-8 or UTF-16, or add an XML declaration that specifies in which character set the markup is encoded.
- Add a DOCTYPE declaration that uses a PUBLIC identifier to identify one of the XHTML DTDs. For more information about XHTML DTDs, see Chapter 8, *Using XML on the Web*.
- Make sure the document has a single root `<html>` element.
- Make all element and attribute names lowercase.
- Add missing end tags such as `</p>` and `</li>`.

- Make sure that elements nest rather than overlap. For example, change:

```
<p><it>an italicized paragraph</p></it>
```

  to

```
<p><it>an italicized paragraph</it></p>
```

- Put double or single straight quotes around attribute values.
- Adjust attributes that have the following form:

```
<input type="checkbox" checked>
```

  so that they have appropriate attribute values:

```
<input type="checkbox" checked="checked">
```

- Replace any occurrences of '&' or '<' in character data within element content or in attribute values with the & and < entity references.
- Change empty elements such as <img> to <img/> or <img></img>.
- Add hyphens to comments so that, for example, <! comment> becomes <!--comment -->.
- Remove nonstandard HTML elements such as <marquee>.
- Add required attributes such as the alt attribute of <img>.
- Move child elements out from inside elements, such as a <blockquote> inside a <p> element.

Migrating from SGML

When you migrate from SGML to XML, you are migrating from an environment in which information is already represented in structured markup. You would expect this to be easier than migrating from unstructured documents and transforming them into structured XML markup. Nevertheless, the process of migrating from large legacy document systems built around SGML can be fairly complex and costly.

As XML is a subset of SGML, the conversion of documents includes handling of SGML markup features not supported in XML. Some issues that must be fixed in documents migrated from SGML include:

- *Case sensitivity.* XML is case sensitive, whereas SGML is not.
- *Tag minimization.* SGML DTD declarations include the tag omission characters (-) and (o) that indicate whether or not start and end tags

are required in an element. Documents migrated from SGML must have all the required start and end tags.

- *Empty elements.* XML requires the extra slash (`/`) at the end of empty elements.
- *Attributes values.* XML requires quotes around attribute values.

There are a number of other changes that must be addressed and strategies to be considered. The pages at `xml.lander.ca/sgml_xml_cs/sgml_xml_cs.html` are a good source of further information.

If you are migrating from SGML to the XML equivalent of the SGML vocabulary, you must convert the DTD as well as the XML instance documents to XML DTD format, or possibly to XML Schema format. For more information about specific details of migrating SGML DTDs to XML, see the pages at `www.xml.com/pub/a/98/07/dtd/index.html`.

Document Style Semantics and Specification Language (DSSSL) is a language for specifying stylesheets for SGML documents. DSSSL can convert SGML and XML to a wide range of formats including XML. Although it is compatible with any SGML application, DSSSL is most often used with DocBook SGML to convert to DocBook XML and other XML languages.

Converting documents from SGML markup to XML-compliant markup can be done using programming, of course, and there is publicly available software. Much of the software that supported SGML before the advent of XML has been ported to support XML. Migration is likely to be easier if you use tools that support SGML as well as XML: for example, XMetal also supports SGML.

Migrating from FrameMaker

If you are using unstructured FrameMaker, migration to XML will obviously be more difficult than it would be from structured FrameMaker, from which you can export to XML as long as you have the appropriate structured application.

Possible approaches to migration are:

- Using a structured application to export to XML, after migrating from an unstructured environment to a structured environment if required
- Writing scripts and programs to convert files to XML

- Using commercial tools to convert files to XML format.

In most cases, considerable clean-up is required after conversion.

Preparing for conversion

Whichever approach you use for migration, you will probably want to prepare your files before conversion to make the conversion easier. For example, you might want to rewrite and reorganize chapter files so that they are more topic-oriented if you are migrating to DITA or S1000D.

If the migration involves moving from an unstructured environment to a structured environment, you should analyse your existing documents to understand their inherent structure, which is important in producing an appropriate EDD. The Adobe publication *Migrating from Unstructured to Structured FrameMaker* provides information about how to perform this analysis.

If your emphasis is changing from material for printing to online material, you will need to consider cross-references of the form 'on page *xx*'. There is a problem with cross-references that use markers: only structured cross-references that use IDs convert correctly when you export to XML.

In a structured environment, images should always be included by reference rather than copied into a document. Also, as mentioned in Chapter 5, *Authoring with XML*, conditional tags can present a problem when exporting to XML: in FrameMaker 8.0 or later you might consider using the filter-by-attribute feature rather than conditional tags.

Migrating from unstructured to structured FrameMaker

To migrate from an unstructured FrameMaker environment to a structured one, you need to create structured elements from FrameMaker formatting components such as paragraph tags, character tags, markers, cross-references and table components. Essentially, you need to create an EDD and the associated formatting template. You can produce the EDD in the following ways:

- Create the entire EDD yourself.
- Create an EDD from an existing DTD or XML Schema.
- In FrameMaker 7.2 and later only, use a conversion rules table to create a rough draft of the EDD that contains basic element definitions and

formatting to match your unstructured template. There is considerable work involved in configuring the conversion table so that the conversion works properly.

- Modify an existing EDD, which might be one of the sample EDDs supplied with FrameMaker.

The Adobe publication *Migrating from Unstructured to Structured FrameMaker* provides information about migration, including how to create a conversion table. Once you have created and tested the conversion table, you can perform a batch migration of unstructured files.

Using a structured application

To export files from a structured FrameMaker application, you need a working structured application and the associated files, including the EDD, DTD, and any read/write rules and XSLT stylesheet. Depending on your version of FrameMaker, you may have the necessary structured application already: DocBook, DITA and S1000D have been supported to at least some degree since Version 7 of FrameMaker. Otherwise, you might need to build your own structured application, which is not an easy task. Refer to Chapter 5, *Authoring with XML* for more information about structured applications and issues with exporting to XML.

If you are migrating from a structured FrameMaker environment and a DTD does not already exist for the target XML language, you can create a DTD from your EDD. You should expect to do some cleaning up of the created DTD.

Using programming and scripting languages

A possible means of converting files to XML, if the necessary programming expertise is available, is to use programming or scripting languages. OmniMark from Stilo International is a programming language with a built-in XML parser that is used mostly in the publishing industry for format conversion. The open-source Perl language is another language used for this purpose, and the FrameScript scripting language is designed to work with FrameMaker to implement FrameMaker publishing solutions.

Programming and scripting languages provide powerful file processing and manipulation capabilities. If you are suitably skilled, you can use them to write programs that provide more control over conversion than commercial tools, possibly resulting in less clean-up work after conversion.

Using commercial tools

A number of commercial tools are available for converting FrameMaker files. These tools include:

- Mif2Go from Omni Systems, which can convert both unstructured and structured files to DITA, DocBook and various other formats. You can convert single files or whole books at once.
- xDoc from Cambridge Docs, which can transform FrameMaker files into DITA XML and DocBook. The DITA and DocBook samples provided include several sample mappings that you can use as the starting point for your conversions to DITA or DocBook.
- Quadralay WebWorks Publisher includes templates that generate XML that can be displayed in browsers that support XSL and CSS.

Migrating from Word

You can export Word documents to XML format: in Word 2003 and later you can save to WordML format, which also features in the Office OpenXML format used in Word 2007. However, both of these essentially just capture the formatting information in a Word document. In Office Professional and stand-alone Word 2003 and later, you can in addition export XML documents for any XML schemas that you have attached to Word.

None of the XML formats you can export are likely to be formats that you want to use. One possibility is to use or develop XSLT stylesheets to transform Word XML formats to something more useful; in Word you can apply such a stylesheet when you save to XML. Another possibility is YAWC Pro (`www.yawcpro.com`), a plug-in for Word that enables you to edit and export Simplified DocBook documents in Word.

There are various commercial tools available for converting from Word to the XML languages you are likely to want to use. For example, the xDoc converter from CambridgeDocs can convert from Word to DITA, DocBook or other DTDs. XMLSpy also has facilities for converting to XML files.

If you use a tool, you will still need to think about and probably adjust the mapping between Word styles and elements in the DTD or schema to which you are migrating.

Migrating between XML languages

To migrate markup documents from one XML language to another you certainly need XSLT stylesheets to convert your source documents. You might also need to restructure your documentation before conversion. For example, if you are migrating from DocBook to DITA and you are creating topic-oriented information in DocBook already, the transition should be fairly easy. On the other hand, if you are creating traditional book-based information in DocBook, there will be additional work involved in breaking your content into topics, not to mentioned the shift in mind-set to topic-based authoring.

7

Transforming XML

This chapter discusses the XML technologies used to transform XML documents to other formats:

- XSLT, the XML language for transforming XML into other formats, including other XML vocabularies
- XPath, the language used to locate parts of an XML document
- XSL-FO, the XML language for describing page layouts.

Due to space limitations, this chapter does not describe these languages in detail: instead it concentrates on the capabilities of the languages and their usage in technical communication.

Extensible Stylesheet Language

To present the content of XML documents for display, you need to transform them into a form that contains the necessary formatting information. For display on the web, you must transform the documents to HTML or XHTML: for display on a printed page, you must transform to a format that contains page formatting information. Extensible Stylesheet Language (XSL) is used for these purposes. It is split into two parts:

- *XSL Transformations* (XSLT). An XML language that specifies rules for transforming XML into another format, the rules being supplied in an XSLT stylesheet that is associated with an XML document. XSLT can transform XML documents to HTML, other XML languages, or indeed any text-based documentation format.
- *XSL Formatting Objects* (XSL-FO). An XML language for describing the layout of text on a page (screen or paper). XSL-FO documents are

usually produced by transformation from XML documents and are used for further transformation to formats such as PDF.

XSLT and XSL-FO are defined by the W3C Recommendations XSLT 1.0, XSLT 2.0 and XSL-FO 1.0. XSL is not the only stylesheet language available: DSSSL was used for similar purposes with SGML and later with XML before XSL was developed - in fact XSL owes much to DSSSL. For more information about DSSSL and supporting tools, see Chapter 4, *XML Documentation Languages.*

The XSL languages are integrated with the publishing tools that technical communicators use. For example, XSLT is used when you produce HTML from XML source documents, while XSLT stylesheets are used in XML editors to show WYSIWYG views of XML documents. You might never need to write or adapt XSLT stylesheets, and indeed XSLT is not an easy language to learn: nevertheless, you can benefit from an appreciation of these technologies and their capabilities.

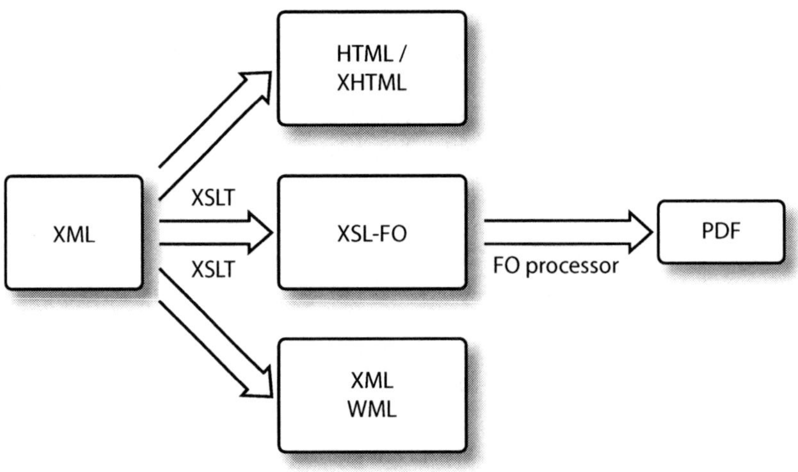

Figure 7.1 XSL transformations

XSLT

While XSLT performs the vital role of styling XML content, it is also a powerful tool for transforming XML documents to other formats. You can:

- Transform one XML vocabulary into another: for example, you can transform DocBook to XHTML, to WML or to XSL-FO for further processing into PDF.
- Transform XML into HTML or XHTML for display in a browser.
- Transform XML into text output. For example, you can transform to RTF format, and XSLT is often used to extract information in the content of XML files into text files.

When you transform XML documents with XSLT, the original document is not changed. You can format a single XML document in different ways by using different stylesheets. For example, you can use separate stylesheets to produce output for the web, for print or for mobile devices, where each stylesheet applies appropriate formatting styles for the specific medium.

XSLT is useful not just for transforming between formats: it has powerful facilities for manipulating the source content and customizing the output. For example, you can:

- Add or remove information to or from the output document
- Rearrange and sort information in the XML document
- Combine multiple input documents into a single output document
- Perform calculations and other functions on the content.

Examples of the uses of XSLT with technical publications include:

- Supplying boilerplate text such as copyright notices, header and footer information
- Selecting particular content according to metadata values
- Hiding text in the source document that you do not want to appear in an output document (conditional processing)
- Generating tables of contents
- Generating indexes with proper sorting
- Generating cross-references and links
- Generating glossaries
- Specifying language-specific rules when transforming documents for different locales (for more information on this, see Chapter 9, *XML and Localization*).

When XSLT transforms and manipulates an XML document, it needs some means of specifying parts or subsets of the document. For example, you might want to locate all elements that have a `platform` attribute with a particular value, or all `<index>` elements. It does this using the XPath language, as described in the following section.

XPath

The language that XSLT and other XML languages use to point to locations in an XML document is XPath (XPath 1.0 and XPath 2.0 are W3C Recommendations). XPath is not an XML language: it uses syntax akin to UNIX path expressions.

XPath uses a data model that represents an XML document as a tree consisting of various nodes for the different types of markup. Each node has information to identify the node and to relate it to other nodes, and each has a value:

- *Root node.* There is a single root node: the value is the value of the document element.
- *Element node.* There is an element node for each element in the document, and each has a list of child nodes, which can include other element nodes, processing instruction nodes, comment nodes and text nodes. The value is the parsed text between the start and end tags of the element.
- *Attribute node.* The value is the attribute value.
- *Text node.* There is a text node for each contiguous run of text within an element node. The value is the text of the node.
- *Processing instruction node.* The value is the data of a processing instruction.
- *Comment node.* The value is the string content of a comment.
- *Namespace node.* The value is a namespace URI.

XPath expressions essentially locate a node or set of nodes and return information about the nodes or their value, which can be a Boolean value, a number or a string. For example, you can locate all nodes corresponding to `<index>` elements, or retrieve the value of an attribute that is the child of a particular element. XPath expresses the relationships between nodes in terms of family relationships that reflect the hierarchy of nodes in the model. This allows you to locate a particular parent element and then specify child elements, sibling elements, ancestors and descendents of that element. You can also perform calculations using XPath as it provides a range of functions.

Note that the XPath model does not allow you to locate everything in an XML document. Entity declarations, CDATA sections and document type declarations are not included in the model. XPath works on the XML document after these items are included: in other words, the tree model applies to a parsed XML document.

How XSLT processing works

To transform an XML document with XSLT, you use an XSLT processor. You might not be aware that you are using an XSLT processor: many publishing tools have integrated XSLT processors, or may use a processor located on a web or application server. Web browsers have built in XSLT processors that allow XML pages with associated XSLT stylesheets to be rendered on web pages. On the other hand, you can also use standalone XSLT processors such as Saxon or Xalan from a command prompt.

An XML document can be considered as a tree containing nodes corresponding to different types of markup or content, such as elements, attributes and text. XSLT can manipulate the markup and content of an XML document by using the XPath language to specify particular nodes or sets of nodes in an XML document. An XSLT stylesheet consists of a set of template rules. Each template rule uses an XPath expression to specify a pattern of nodes and a template to be output when the pattern is matched. The template can include markup, content from the input document and new content.

The XSLT processor works by reading the XML document and applying the set of template rules from the XSLT stylesheet to build the output document. The XSLT processor examines each node of the XML document in turn and compares it with the pattern of each template rule in the stylesheet. When there is a match, it writes whatever is in a rule's template to the output document.

The following section provides an example that illustrates just a few of the elements in XSLT. XSLT and XPath together provide an extensive set of elements for manipulating and transforming XML documents. For example, the `<xsl:import>` and `<xsl:include>` elements enable you to merge multiple stylesheets so that you can combine and reuse stylesheets for different purposes.

A simple XSLT example

This section contains a very simple example of an XSLT stylesheet that transforms an XML document containing personnel information into HTML. The output HTML document displays the names of people, together with their personnel number and their photograph. A snippet of the XML document is as follows:

```
<?xml version="1.0"?>

<?xml-stylesheet type="text/xsl" href="personnel.xslt"?>

<employee staffid="A1956">
   <forename>Jo</forename>
   <surname>Bloggs</surname>
   <image source="jo_bloggs"/>
</employee>
```

Figure 7.2 is the example XSLT stylesheet, which is associated with the XML document through the processing instruction. In the example, the elements of the XSLT and XPath languages are shown in bold.

Figure 7.2 Example XSLT stylesheet

```
<?xml version="1.0"?>

<xsl:stylesheet version="1.0"
    xmlns:xsl="http://www.w3.org/1999/XSL/Transform">

<xsl:template match="/">
  <html>
    <head><title>Employees</title></head>
    <body>
      <h1>Employees</h1>
      <hr>
      <xsl:apply-templates/>
    </body>
  </html>
</xsl:template>

<xsl:template match="employee">
<h2>
<xsl:value-of select="surname"/>, <xsl:value-of select="forename"/> </h2>
<h2>
Personnel Number: <xsl:value-of select="@staffid"/>
</h2>
<img src="{image/@source}.jpg"/>
<hr>
</xsl:template>

</xsl:stylesheet>
```

The XSLT stylesheet produces HTML such as the following:

Figure 7.3 HTML example

```
<html>
  <head><title>Employees</title></head>
  <body>
    <h1>Employees</h1>
    <hr>
    ...
    <h2>Bloggs, Jo</h2>
    <h2>
      Personnel Number: A1956
    </h2>
    <img src="jo_bloggs.jpg"/>
    <hr>
    ...
  </body>
</html>
```

In the stylesheet, the `<xsl:template>` elements contain template rules. The `<xsl:template match="/">` element instructs the XSLT processor to start at the root element of the XML document. The template for `<xsl:template match="/">` specifies that HTML markup (`<html>`, `<head>`, `<body>` elements) is to be output surrounding whatever is output as a result of the `<xsl:apply-templates>` element. The latter element is an instruction to apply any other templates in the stylesheet at this point.

In the example stylesheet, the only other template is that specified in the `<xsl:template match="employee">` element. In the `<xsl:template match="employee">` element, the `match` attribute specifies that all employee elements are to be matched. The template then specifies that various HTML markup is to be output: `<h2>`, `<img>` and `<hr>` elements. It also outputs content from the selected child nodes of <employee>:

- The value of the `<surname>` child element, specified by `<xsl:value-of select="surname"/>`.
- The value of the `<forename>` child element, specified by `<xsl:value-of select="forename"/>`.
- The value of the `staffid` attribute of the `<employee>` element, specified by `<xsl:value-of select="@staffid"/>`.

  Here the XPath expression `@staffid` points to the value of the `staffid` attribute.
- The value of the `source` attribute of the `<image>` child element, specified by `<img src="{image/@source}.jpg"/>`.

Here the value of the HTML `src` attribute value is specified by the XPath expression, which points to the value of the `source` attribute of the `<image>` element in the XML document.

XSL-FO

XSL Formatting Objects (XSL-FO) is an XML application that describes page layouts. An XSL-FO document contains formatting instructions that describe how the content of the document should be laid out.

XSL-FO documents are not usually authored directly. An XML document in XHTML, DocBook, DITA or any other XML language is first transformed to XSL-FO format using XSLT, then an *FO processor* converts that document into a printable or readable format, usually PostScript or PDF, although formats such as RTF, T_EX or other print-ready forms can also be produced.

The FO processors available range from the free Apache FOP up to expensive processors. FO processors may be stand-alone or built into your publishing software. Processors vary in the output formats that they can create, and in their degree of support for the XSL-FO specification. Although there are currently no web browsers that can display documents marked up with XSL-FO directly, this might be possible in the future.

Needless to say, people who produce XSL-FO documents and related stylesheets require a number of skills, including an appreciation of page layout, as well as XML knowledge, particularly a grasp of how XPath and XSLT are used to select and reorganize XML source content. XSLT stylesheets for transforming DocBook to XSL-FO are available: see `docbook.sourceforge.net`.

The XSL-FO language

The root element of an XSL-FO document is `<fo:root>`, which contains:

- An `<fo:layout-master-set>` element, containing templates for the pages to be created
- One or more `<fo:page-sequence>` elements, containing the text and images to be placed on the pages.

When an FO processor reads an XSL-FO document, it creates a page based on the first template specified by the `<fo:layout-master-set>` element.

It then fills the page with content from the associated `<fo:page-sequence>` element. After filling the first page, the FO processor creates a second page based on a template and fills it with content, and the process continues until there is no more content.

Page master templates can define the size of a page and the margins of each page. More importantly, they can define sequences of pages in which the odd and even pages are laid out differently. They can also define the directions for the flow of text, which is essential for documents containing bidirectional languages such as Arabic and Hebrew, which flow from right to left by default. When, for example, English words are embedded within sections of these bidirectional languages, the direction of flow must change from left to right.

Within an `<fo:page-sequence>` element, the content of an XSL-FO document is stored in several kinds of elements, which are descendants of either an `<fo:flow>` or an `<fo:static-content>` element. The types of elements are:

- Block-level formatting objects
- Inline formatting objects
- Table formatting objects
- Out-of-line formatting objects
- List objects
- Link objects
- External objects.

Typically there is a sequence of *flows* (contained within the `<xsl:flow>` or `<xsl:static-content>` element) in which each flow is attached to a page layout. The flows contain a list of blocks which, in turn, each contain a list of text data, inline markup elements, or a combination of the two. There can also be content added to the margins of the document, for page numbers, chapter headings, and so on.

The content of an XSL-FO document is mostly text, but it can contain non-XML content such as GIF and JPEG images by including them in a similar way to that of the `<IMG>` element in HTML. Other forms of XML content, such as MathML and SVG, can be embedded directly inside the XSL-FO document.

A simple XSL-FO example

This section contains a very simple example of a document marked up using XSL-FO. The root element contains one `<fo:layout-master-set>` and one `<fo:page-sequence>` element. The `<fo:layout-master-set>` element contains one `<fo:simplepage-master>` child element, which describes the type of page on which content is to be placed. In this example there is only one very simple page, but typically documents have different master pages for first, right and left body pages, front matter, back matter and so on. Each of the master pages can have different margins, page numbering and other properties.

The `<fo:page-sequence>` element specifies the content to be placed in a copy of a master page, while the `master-name` attribute specifies the master page to be used. The `<fo:flow>` child element holds the actual content to be placed on the pages. In this example, the content comprises two `<fo:block>` objects, each with a `font-size` of 10 points, a `font-family` of `serif` and a `line-height` of 20 points.

```
<?xml version="1.0"?>

<fo:root xmlns:fo="http://www.w3.org/XSL/Format/1.0">

  <fo:layout-master-set>
    <fo:simple-page-master page-master-name="only">
      <fo:region-body/>
    </fo:simple-page-master>
  </fo:layout-master-set>

  <fo:page-sequence master-name="only">
  <fo:flow flow-name="xsl-region-body">
      <fo:block font-size="10pt" font-family="serif" line-height="20pt">
        Peter
      </fo:block>
      <fo:block font-size="10pt" font-family="serif" line-height="20pt">
        Wrightwell
      </fo:block>
    </fo:flow>
  </fo:page-sequence>

</fo:root>
```

8

Using XML on the Web

This chapter describes how XML can be used on the World Wide Web:

- As XHTML, the XML-compatible version of HTML
- As XML styled with a Cascading Style Sheet (CSS)
- As XML styled with an XSLT stylesheet.

For each of these methods, information about Web browser support is given.

XML and the web

One of the original intentions for XML was as a markup language for the web, a standard language that would overcome the problem of the inconsistent versions of HTML used in different web browsers. If you simply open an XML document with a web browser such as Microsoft Internet Explorer (IE), you see something like Figure 8.1:

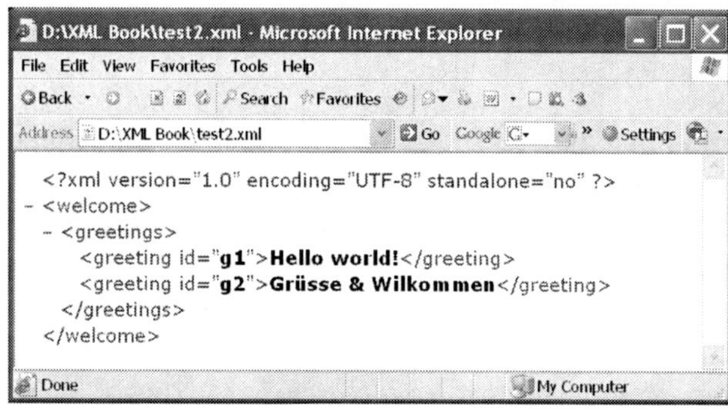

Figure 8.1 XML document displayed in Internet Explorer

The browser simply shows a structural view of the XML file, which is not surprising when you consider XML's separation of content from formatting information. So how can a web browser understand how to display an XML document if there is no fixed set of elements in XML and no formatting information? There are two ways in which you can use XML documents on the web, albeit within the limits of what is supported by particular browsers:

- As Extensible HyperText Markup Language (XHTML), the XML equivalent of HTML. You can either create XHTML documents directly, or transform XML documents into XHTML for display on the web.
- In XML pages that have a stylesheet. An XML document can have a CSS or an XSLT stylesheet that defines how the content of elements should be displayed in a web page when the XML is rendered to HTML.

For displaying XML documents with stylesheets, browsers have built-in software for processing the stylesheet and applying the formatting. For example, Internet Explorer has an XML parser that applies the XSLT stylesheet to transform the XML to HTML, which is what is actually displayed. Scripting languages such as JavaScript can also be used with XML documents, to manipulate data as well as enhance the interactivity of web pages.

Browser support for XML

To see anything of an XML document at all in a browser, the browser must be XML-enabled. Fortunately this includes all more recent browsers, for example Internet Explorer Version 5 and later. An XML-enabled browser checks for well-formedness, although it does not usually check for validity. As shown in Figure 8.1, an XML document without a stylesheet displays in Internet Explorer as a structure view in which (depending on the browser version) you can click on the + and – signs to expand or collapse the structure.

When you use XML with CSS or XSLT stylesheets, what you see in a web browser view depends very much on the browser's level of support for those technologies. Different results can be obtained from different browsers because, for example, one browser supports one subset of the CSS specification and another a different subset of the specification. Also, there are different versions of CSS and XSLT, so a particular browser might support one version of these technologies but not another. All this means that there are major differences in the support for XML between, for example, Internet Explorer Version 5 and Internet Explorer Version 6, as well as between different browsers. The support for XHTML by browsers is also inconsistent – for more information, see *Browser support for XHTML* on page 161.

It is impossible to discuss the XML support for every web browser version in this book, but more information is given in following sections. The website at `www.webdevout.net` provides detailed information about browser support of CSS, HTML and XHTML. Support for Scalable Vector Graphics (SVG) in web browsers is discussed in Chapter 10, *Scalable Vector Graphics.*

XHTML

XHTML was developed by the W3C as a standard extensible markup language for web pages that is based on XML rather than SGML. XHTML has similar elements to HTML, but XHTML documents must be well-formed, so XHTML differs from HTML in the following ways:

- Elements must have end tags and must be nested properly
- Attribute values must be inside single quotes
- Empty elements must end in '/>' instead of just '>'
- XHTML is case sensitive.

XHTML 1.0 became a W3C Recommendation in 2000 and XHTML 1.1 in 2001. There are actually three different `DOCTYPES` for XHTML 1.0, and each has elements equivalent to respective HTML 4.01 versions, as Table 8.1 shows.

All XHTML elements must be in the appropriate XML namespace for the version being used. This is usually done by declaring a default namespace on the root element using the `xmlns` attribute, as in the following example for XHTML 1.0 and XHTML 1.1:

```
<html xmlns="http://www.w3.org/1999/xhtml">
```

Table 8.1 XHTML versions

Version	Doctype	Notes
XHTML 1.0 Strict	`<!DOCTYPE html PUBLIC "-//W3C//DTD XHTML 1.0 Strict//EN" "http://www.w3.org/TR/xhtml1/DTD/xhtml1-strict.dtd">`	Equivalent to HTML 4.01 Strict, but follows XML syntax rules.
XHTML 1.0 Transitional	`<!DOCTYPE html PUBLIC "-//W3C//DTD XHTML 1.0 Transitional//EN" "http://www.w3.org/TR/xhtml1/DTD/xhtml1-transitional.dtd">`	Adds a number of elements to XHTML 1.0 Strict.
XHTML 1.0 Frameset	`<!DOCTYPE html PUBLIC "-//W3C//DTD XHTML 1.0 Frameset//EN" "http://www.w3.org/TR/xhtml1/DTD/xhtml1-frameset.dtd">`	Allows definition of HTML framesets.
XHTML 1.1	`<!DOCTYPE html PUBLIC "-//W3C//DTD XHTML 1.1//EN" "http://www.w3.org/TR/xhtml11/DTD/xhtml11.dtd">`	Module-based XHTML, which is a reformulation of XHTML 1.0 Strict.
XHTML 2.0	`<!DOCTYPE html PUBLIC "-//W3C//DTD XHTML 2.0//EN" "http://www.w3.org/MarkUp/DTD/xhtml2.dtd">`	The current working version as of early 2010.

There are also versions of the XHTML family designed for specific devices or uses:

- *XHTML Basic.* For use in hand-held devices and mobile phones and intended to replace WML.
- *XHTML Mobile Profile.* For hand telephones: adds mobile phone-specific elements to XHTML Basic.
- *XHTML+Voice.* For visual and voice interactions.

XHTML provides a number of advantages over HTML. For example, you can use XML tools to process XHTML, and it is easily converted to other XML languages such as WML. As mentioned, you can also use versions of XHTML on a range of output devices, such as hand-held and mobile devices, and in specialized browsers, such as speech-enabled browsers. Furthermore, you can mix other XML vocabularies such as SVG and MathML into an XHTML document (as long as the browser can understand those languages). However, the advantages of XHTML have yet to be fully realized due to incomplete browser support, as described in the following section.

Browser support for XHTML

No browser fully supports XHTML yet. Also, there are various other inconsistencies in the way in which browsers handle some common XHTML constructs.

One problem lies in the way in which web servers handle XHTML files. When a website sends a document to your browser, it adds a special content type header that tells the browser what kind of document it is. For example, web servers add the content type `text/html` to HTML documents for files of type `.htm` or `.html`. The proper content type for XHTML is actually `application/xhtml+xml`. However, many web servers do not have this content type reserved for any file extension, so cannot automatically add the content type to XHTML files. Instead, they add the `text/html` content type to the XHTML file. The browser therefore assumes that the document is actually an HTML file. This means the advantages of XHTML are lost: for example, SVG and MathML incorporated in the document are not understood.

Even Internet Explorer Version 7.0 does not really support XHTML documents, because it does not recognize the `application/xhtml+xml` content type. It only renders XHTML documents correctly when they are served with the `text/html` content type and authored according to HTML compatibility guidelines.

Another reason for the slow adoption of XHTML (and CSS2) is the fact that many web pages have been generated using proprietary software, such as Microsoft FrontPage, that produce non-standard HTML by default. This has maintained the support by browsers for non-standard HTML at the expense of standards such as XHTML. This, and the fact that XHTML 2.0 is not backwards compatible, has led to considerable debate amongst the web development community as to whether XHTML should be adopted at all.

Cascading Style Sheets

You can use Cascading Style Sheets (CSS) for styling XML in browsers or for styling HTML generated from XML. CSS was introduced in 1996 as a standard for adding information to HTML documents about style properties such as fonts and borders. Later CSS became compatible with XML as well.

You can include CSS statements directly in an XHTML document, but the best practice is to use a separate stylesheet document with the extension `.css`. You can prepare a stylesheet in any text editor, as stylesheets are merely text files. Some dedicated CSS editors are also available, and XML editors generally support CSS.

To associate an XML document with a CSS stylesheet, you use a processing instruction such as the following:

```
<?xml-stylesheet type="text/css" href="project.css"?>
```

This example is for a stylesheet located in the same folder as the XML document, but you can also specify a relative file path or a URL to locate a stylesheet, for example:

```
<?xml-stylesheet type="text/css"
href="http://www.acme.org/xml/styles/project.css"?>
```

You can use the same stylesheet for many documents: typically they are located in a central location, possibly on a web server where all of your documents can refer to them.

Note that you can style XML documents with either CSS or XSLT, or use a combination of XSLT and CSS. However, if you need to restructure documents by adding or removing information, or by rearranging and sorting, you must use XSLT.

CSS syntax

There are currently three levels of CSS:

- *CSS level 1.* The original HTML-only version of CSS and a W3C Recommendation.
- *CSS level 2.* A version that added a number of style properties and support for XML. Level 2 is a W3C Candidate Recommendation. CSS 2.1 corrected a few errors in CSS level 2.
- *CSS level 3.* The latest work-in-progress version.

A CSS stylesheet contains a list of rules. Each rule consists of a selector, which is the name of the element or elements in the XML document to which the rule applies, followed by the style properties to apply to the content of the elements.

In the example below, the selector in the first rule is for the `<element1>` element and the rule specifies that the contents of the element should be displayed in a block by itself (`display: block`). The second rule specifies that the contents of the `<element2>` element should be displayed in a block by itself (`display:block`) in 12-point (`font-size: 12pt`) and in bold type (`font-weight: bold`).

```
element1 { display: block }
element2 { display: block; font-size: 12pt; font-weight: bold }
element3 { display: block; margin-bottom: 10px }
```

In CSS, the style properties you can specify include font properties (family, size, weight), colour properties of items, text properties and properties governing the layout of element content on the page.

Different browsers render CSS layouts differently. Earlier versions of the major browsers did not support CSS for XML documents at all, while later versions, such as Opera 4.0 and 5.0 and Internet Explorer versions 5.0 and 5.5, all implemented some but not all parts of the CSS specification. To complicate matters further, browsers sometimes do not implement the same subsets of CSS for XML as they do for HTML. The major browsers have been slow to support the CSS 2.1 specification, although later versions of Internet Explorer and Firefox support most if not all of the specification.

One of the uses of XSLT is for transforming XML into HTML or XHTML for display on the web.

Many browsers support client-side XSLT: in other words, they have a built-in XSLT processor. Another possibility is for the transformation to HTML to be done at the web server.

To associate an XML document with an XSLT stylesheet, you use a processing instruction such as the following:

```
<?xml-stylesheet type="text/xsl" href="project.xslt"?>
```

As with CSS, you can use the same stylesheet for many documents, so typically XSLT stylesheets are located in a central location. You can also use XSLT and CSS together. For example, you could use XSLT to transform an XML document into XHTML, which then uses a CSS to define how it is rendered. As described in Chapter 7, *Transforming XML*, an XSLT stylesheet is effectively a set of template rules that determine the output generated in the transformation. Essentially, the XML document's content is 'poured' into a template of HTML or XHTML tags.

XSLT 1.0 became a W3C Recommendation in 1999, while XSLT 2.0 became a W3C recommendation in 2007.

Browser support for XSLT

Browser support for XSLT varies: different browsers render the same XML document differently. You should beware of using Internet Explorer Version 5 in particular, as it supports Microsoft's own version of XSLT (based on an early W3C specification) and not the official XSLT 1.0 specification. A stylesheet that uses Microsoft's custom version of XSLT has the namespace:

```
http:www.w3.org/TR/WD-xsl
```

while a stylesheet confirming to the XSLT 1.0 specification has the namespace:

```
http:www.w3.org/1999/XSL/Transform
```

From Version 6, Internet Explorer supports the XSLT 1.0 specification.

9

XML and Localization

This chapter discusses the uses of XML for localization, including:

- XML's advantages for localization, including the reason for those advantages and the cost-saving benefits that they provide.
- The ways in which XML is used in the localization of documentation; both as the file format for material that is translated and for the XML languages that are used to help with the localization process.
- How your use of XML as a source format can make translation easier.

Overview

Localization is the translation and adaptation of a product to different cultural conventions. It involves not only the translation of technical publications from English to a number of other languages, but the translation of all localizable material in software products, such as user interface strings, icons, error messages and other text.

Localization is an important business: huge revenue can be generated by organizations when they localize their products. Nevertheless, translation is very costly, so organizations are continually looking for ways to reduce translation cost and timescales to realize profits sooner. Luckily, XML offers many advantages for reducing cost and time to market, as discussed in the following section.

Advantages of XML for localization

XML has the following advantages for localization:

- A character encoding system that allows an XML document to contain characters in virtually any language.
- Separation of content from formatting, which allows single-source publishing, which in turn simplifies the localization process and reduces costs.
- A rich set of supporting tools for validating and transforming XML documents, which means that many translation processes can be automated and various translation challenges are easier to overcome.
- Powerful facilities for reuse, which reduces word count and therefore translation cost.
- Facilities for making translation easier, such as differentiating translatable and non-translatable content and identifying terminology.
- A standard format that allows easy document interchange between XML-compliant tools.

Character encoding in XML

The content and markup of XML documents is not restricted to ASCII characters: it can contain characters from almost any language. You can therefore provide an XML file in English for translation and the translators can return an XML file with content in any language. For example, the following is an excerpt of an XML document with Japanese content:

```
<?xml version="1.0" encoding="utf-8" ?>

<p>
高度な検索を使用すると、次の操作が行えます
</p>

</xml>
```

XML's handling of languages is due to its support for Unicode, a character encoding system that embraces practically all of the world's languages. Of course, to enter Unicode characters in an XML document, you need an editor that supports Unicode: there are many that do so. You also need the necessary fonts for displaying the language scripts.

To use the Unicode system, an encoding scheme is required to specify the digital representation of each character. Unicode Transformation Format (UTF) is such an encoding scheme for Unicode: UTF-8 and UTF-16 are two variants used in XML documents. UTF-8, a superset of ASCII, is the default encoding for XML documents: if you do not include an encoding attribute in the XML declaration of your XML documents, UTF-8 encoding is assumed.

```
<?xml version="1.0" encoding="utf-8" ?>
```

When content created using a traditional WYSIWYG authoring environment is localized, as shown in Figure 9.1, there can be considerable overhead in removing formatting constructs before translation, and there is also the considerable time and expense involved in publishing the documents for each of the output formats for the translated content. Often many manual corrections are required for the formatted localized content before it is returned by the translators.

In an XML authoring environment, as shown in Figure 9.2, content is separate from formatting: the formatting data typically is contained in one or more associated stylesheets that can be localized separately from the content data. The localized content is returned by the translators, and the output formats are generated from a single source. You only translate content once rather than once per output format. The benefits are quicker translation, as there are fewer formatting constructs to handle, as well as improved *translation memory* usability and a reduced need for costly and tedious corrections to the formatted localized content.

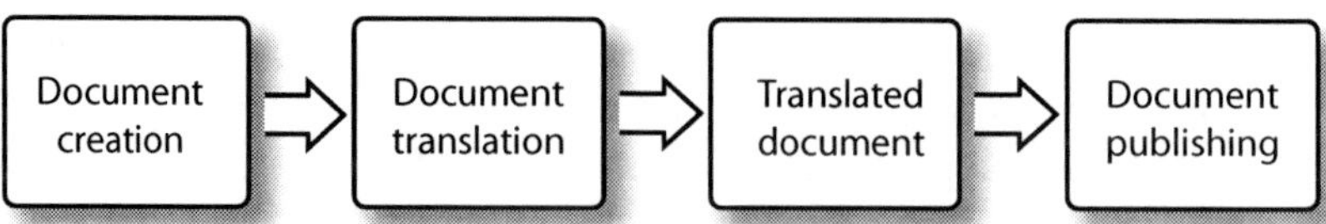

Figure 9.2 XML translation workflow

XML and automation of localization processes

By using XSLT stylesheets to transform XML documents, you can automate processes for overcoming some of the challenges of translating documentation. If XML is used as a source format, you can transform the same content by using XSLT stylesheets tailored for different languages and locales.

For example, the handling of numerical data varies with locale. The comma is used as the decimal point character in some numbering systems, and some languages do not use the comma to separate thousands, or separate numbers in different groupings. To handle these differences, you can use the XSLT element `<xsl:number>` and the function `format-number()` to format numerical values during transformation of XML documents. For example `format-number(1234.56, "#,###.##")` generates '1,234.56'.

The rules for sorting also differ between languages: for example, in German, lowercase letters are sorted before uppercase letters. In fact, not all languages have the concept of alphabetical order: ideographic languages such as Japanese do not. In XSLT, the element `<xsl:sort>` allows you to specify how the result of an `<xsl:apply-templates>` or `<xsl:for-each>` element is sorted for a particular language.

Some other translation challenges can be handled by using CSS. For example, CSS can handle bidirectional languages such as Arabic and Hebrew, which are written right to left, top to bottom unless left-to-right inclusions are made.

Also, text formatting such as bold, italic, underline, and so on might need to be different depending on the language of the translated text. The following example shows how the `lang()` selector of CSS-2 can be used to specify different formatting for an `<important>` element for different languages:

```
important:lang(en) { font-weight: bold; }
important:lang(fr) { font-style: italic; }
```

To render XML documents with the appropriate style for a particular language, you must identify the language of the content: you can use the `xml:lang` reserved attribute for this purpose. For example, for German content:

```
<p xml:lang="de">Text auf Deutsch.</p>
```

The values for the `xml:lang` attributes can be any of the language tags defined in RFC 3066 *Tags for the Identification of Languages*: see `www.faqs.org/rfcs/rfc3066.html`.

You can identify the language for the whole document or part of a document: the `xml:lang` attribute applies to all attributes and content of the element where it appears and all children of that element, except when overwritten. Although you can have XML documents that contain a mixture of languages, they are not currently easy to localize, so their use is not recommended.

If you use a DTD and validate your documents, you must declare `xml:lang` in the DTD as for any other attribute:

```
<!ATTLIST para xml:lang NMTOKEN>
```

In XML documents based on XML Schema, you can use a `lang` attribute in the same way as `xml:lang`. The `lang` attribute has the built-in simple type `xs:language`, which constrains its value to one of the international language codes defined by RFC 1766: see `www.faqs.org/rfcs/rfc1766.html`.

Lowering the cost of translation through reuse

The cost of translation is related to word count. Fortunately, XML has various reuse facilities for reducing the number of words that need to be translated, and thereby the cost of translation:

- With XML documents based on a DTD, you can use general entities for boilerplate text or to include external reused material, and there are similar facilities for reuse in the XML Schema language.

- The structured authoring that XML enforces with its modules of content makes reuse easier.
- DITA has many facilities for reuse, including the `conref` mechanism discussed in Chapter 4, *XML Documentation Languages.*
- If you reuse strings from the user interface of a product in the documentation, this means you only need to translate those strings once.
- As XML elements can correspond to segments in translation memory, the use of the translation memory, which is itself a powerful reuse mechanism, is more efficient. Translation memory is discussed further in *XML in the localization process* on page 172.

The cost savings from reduced word count are compounded for each language into which the documentation is translated.

One way in which XML can be useful to translators is in providing elements and attributes that guide the translator, for example, in identifying:

- Content that is translatable or not translatable (such as programming commands)
- Terms and their context
- Notes to translators about how to translate particular content
- Elements with content that needs to be translated according to specific rules.

Such facilities have been used in XML vocabularies for some years: for example, DITA has a `translate` attribute, but a standard set of localization-related elements and attributes would obviously be beneficial. The Internationalization Tag Set (ITS) 1.0, a W3C Recommendation developed by the W3C Internationalization ITS Working Group, addresses this need for localization directives.

The ITS language provides elements and attributes that you can use in any XML document to specify localization-related information. The following examples illustrate some of the ITS markup. To distinguish non-translatable text within translatable content, you can use an `its:translate` attribute:

```
<p id="101">Our company, <ph its:translate="no">TechDoc
Solutions Ltd.</ph> </p>
```

To delimit terms, you can use an `its:term` attribute:

```
<h1><span its:term="yes">Document Management</span></h1>
```

You can also use the optional `its:termInfoRef` attribute whose value contains a URI that refers to a resource that provides contextual information about the term: such contextual information increases translation accuracy.

To provide specific instructions to translators, you can use an `its:locNote` attribute, for example:

```
<data its:locNote="Note to translator" Text to be translated
</data>
```

Rather than use attributes, you can also specify global translation rules in the header of XML documents, where you use an `<its:rules>` element containing child elements specifying rules for which elements contain translatable content, which elements should be translated according to specific rules, and so on.

You must add any `its` attributes that you use to your XML schema document. For example, to declare an `its:translate` attribute for a `<book>` element in a DTD:

```
<!ELEMENT book … >
<!ATTLIST book its:translate (yes|no) #IMPLIED>
```

You would probably want to declare such an attribute for all elements. You can also use `its` in XML languages that already have similar facilities. For example, you can associate the `its:translate` attribute with the DITA `translate` attribute. Admittedly, ITS does not yet address all the issues affecting translators: for example, it does not address the issue of handling elements with preformatted content in which white space is significant.

Another area that has been troublesome for translators is the translation of graphics. The 'text' in bitmapped formats such as GIF and JPEG is not editable, and translators need access to graphics tools if they are to edit graphics source files. However, graphics coded in the XML Scalable Vector Graphics (SVG) language allow easy access to text for translation. This is much easier than handling other graphic types with embedded text.

Easier document interchange

XML allows for easy interchange of data. This is important for localization for the following reasons:

- XML languages can be used for exchanging data between translation tools and between organizations.
- Localization includes the translation of strings in user interfaces of software applications. Often these strings are encoded in XML files and are easily transformed to the XML languages used for translation.

XML in the localization process

XML is not only useful as a source format in localization: as Figure 9.3 shows, XML languages play an important role in each of the phases of a typical translation process - preprocessing, translation and postprocessing. The boxes in the centre of the diagram represent the tools used for prelocalization preparation and for translation, tools that use translation memory and terminology management. Often localization workbenches combine all these translation technologies into a single application.

The Localization Industry Standards Association (LISA), specifically the Open Standards for Container/Content Allowing Reuse (OSCAR) working committee, has been responsible for developing many of the XML languages used in localization, including TMX, TBX and SRX, as standards. OASIS also has contributed to XML-based standards for localization through its endorsement of XLIFF (XML Localization Interchange File Format).

If you use XML as a source format for your documentation, translators can use the XML directly. If XML is not the source format, the translatable text can be extracted from a variety of source formats into a standard XML format, XLIFF (or OpenTag, a similar language). XLIFF provides a common format for exchanging localization data between software organizations and localization vendors, or between localization tools such as translation memory systems and machine translation systems.

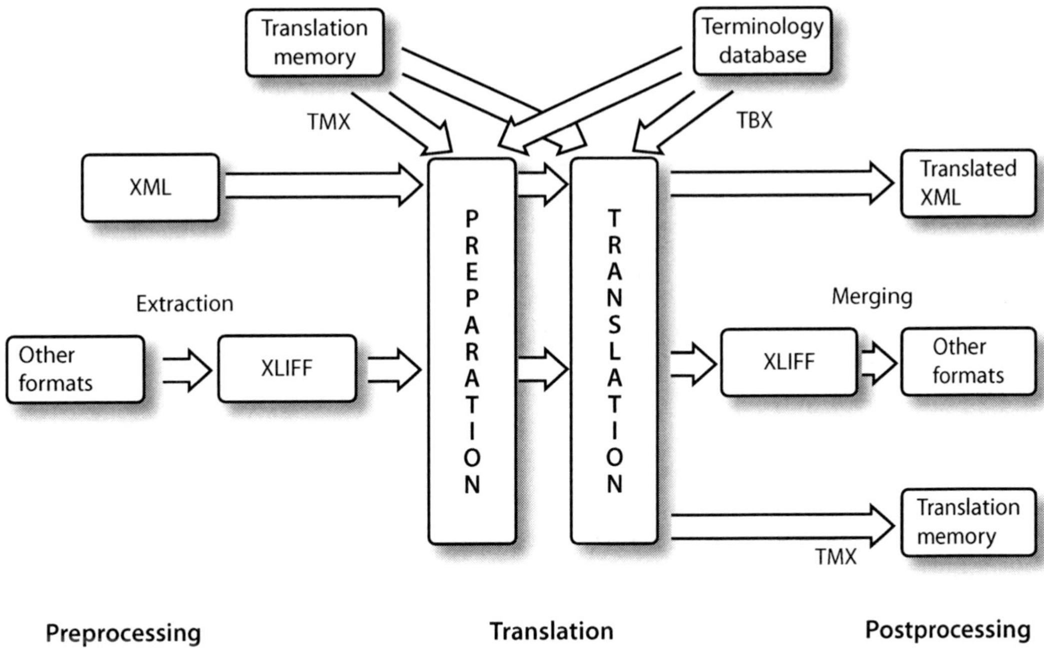

Figure 9.3 XML and its role in localization

The extraction is done by computer aided translation (CAT) tools, which have filters to separate the text and formatting for common formats such as HTML and RTF. After an XLIFF document is translated, the translated strings are extracted and re-inserted into the original document in a process called *reverse conversion* or *merging*.

Process automation based on standards such as XLIFF has brought down the cost of localization considerably and made possible the localization of material until recently considered unsuitable for it.

An XLIFF document contains a collection of translation units represented by `<trans-unit>` elements. Each translation unit consists of a `<source>` element containing a sentence or paragraph extracted from the original document, and a `<target>` element, which the translator has to complete with the translation. If the material has been translated for previous releases, the document can also contain `<alt-trans>` elements that contain possible translations extracted from a translation memory (TM) database in the preparation phase: translators can use these translations as a guideline.

The following is an example excerpt from an XLIFF document:

```
<trans-unit approved="no" id="1" resname="res1" xml:space="preserve">
   <source xml:lang="en">technical communication</source>
   <target xml:lang="fr">communication technique</target>
   <alt-trans match-quality="100" origin="web" tool="TM Search">
      <source xml:lang="en">technical communication</source>
      <target xml:lang="fr">communication technique</target>
   </alt-trans>
   <alt-trans match-quality="92" origin="web" tool="TM Search">
      <source xml:lang="en">"technical communication"</source>
      <target xml:lang="fr">"communication technique"</target>
   </alt-trans>
</trans-unit>
```

In the translation phase, the files to be translated are processed by tools such as translation memories and machine translation systems. Machine translation systems are used for automated translation of documentation not previously translated. On the other hand, for documentation where previous releases have been translated, translation memories allow the translator to identify text that is already translated. This reuse of translated material of course reduces translation cost.

The translator uses tools to provide all the missing translations and to verify those provided at the prelocalization preparation stage. Translation tools (editors) handle source files in pieces called *segments*, which are usually sentences. Translation memories work by looking up segments in a database of previously translated segments and their translations. Segments from the database that match the input segment exactly, or segments that are similar to the input segment, are presented to the translator as suggested translations.

Storing data in smaller topic-sized chunks, rather than book chapters, makes it easier to pinpoint changes and to update translations quickly, making more efficient use of translation memory. Individual small chunks are less likely to change after translation. meaning more efficient use of translation memory.

Translation Memory eXchange (TMX) is a standard language for the exchange of translation memory data created by localization tools. It has been adopted by the localization community as the best way of importing and exporting translation memories. After translation, the TM database is updated with newly translated content, which is stored in TMX format.

A TMX document is similar to an XLIFF document, in that it contains a collection of translation units, in this case represented by the <tu> element. Each <tu> element contains at least two translation unit variants, <tuv> elements, which specify text in a given language. The following example shows the candidate translations used in the XLIFF example:

```
<tu tuid="1292672451312" creationdate="20070112T212503Z">
   <tuv xml:lang="en">
      <seg> technical communication</seg>
   </tuv>
   <tuv xml:lang="fr">
      <seg> communication technique</seg>
   </tuv>
</tu>
```

Another XML language, Segmentation Rules eXchange (SRX) is used for defining how TM tools segment source texts to produce their databases, which is important in making TM data portable between tools. SRX is intended to enhance the TMX standard so that TM data exchanged between applications can be used more effectively.

The use of consistent and correct terminology reduces translation costs by making it easier for the translator to locate the appropriate translations for terms and by increasing the reuse of translated terms. So that terms can be translated consistently, terminology management systems store terms and their translations in terminology databases. The TermBase eXchange (TBX) language is a LISA standard for the interchange of terminological data with terminology databases.

Another XML standard adopted by OSCAR in 2007 is xml:tm, a language for embedding *text memory* within XML documents. Text memory comprises author memory and translation memory. Author memory is used to track changes during the authoring cycle, by means of unique identifiers for each text unit within a document. Translation memory uses the information and identifiers from author memory to implement translation memory matching. xml:tm can be used on any document that can be converted to XLIFF, not just XML documents.

Global Information Management Metrics Exchange (GMX) is yet another OSCAR standard, a family of standards for globalisation- and localisation-related metrics. One of these standards is GMX-V, which is concerned with word counts and other statistics used to quantify localization and translation workload. GMX-V markup can be used within XLIFF documents or within stand-alone files.

The OASIS and OSCAR standards covered in this chapter, together with ITS, Unicode, DITA, DocBook and other XML-related standards, are embraced by the Open Architecture for XML Authoring and Localization (OAXAL) model. This is an OASIS initiative to encourage the development of an open standards approach to XML authoring and localisation

Although most localization tools, for example SDL International's SDLX and TagEditor from TRADOS, support XML, not all of them provide perfect handling of XML. Some have various problems that, depending on how your XML documents are set and which XML features they use, may make the localization process more difficult.

Making XML documents easier to localize

When you author XML documents, there are best practices that you can follow to make the documents easier to translate. The Internationalization Tag Set (ITS) Working Group of the W3C Internationalization Activity has published a *Working Draft of Best Practices for XML Internationalization* document. The document contains guidelines for authors of XML documents and developers of XML applications who are producing or customizing XML schema documents. Some of the main guidelines are:

- Use xml:lang to identify the language of content.
- Avoid using CDATA notation in translatable XML content.
- Avoid the use of attribute values to contain translatable data (for XML schema authors).
- Name elements according to their purpose, not according to the rendering of their content (for XML schema authors). For example, if an element is used to emphasize text, call it `<emph>`, not `<bold>`.

Unsurprisingly, the guidelines suggest using ITS to achieve many of the best practices.

10

Scalable Vector Graphics

This chapter:

- Discusses Scalable Vector Graphics (SVG), the XML standard for describing vector graphics.
- Describes the advantages of SVG for technical documentation.
- Discusses the different ways of generating and viewing SVG markup.
- Summarizes the support for SVG in web browsers and publishing tools.

Introduction

Scalable Vector Graphics (SVG) is an XML application for describing two-dimensional line art. An SVG document contains a definition of a vector graphic and elements that represent polygons, rectangles, ellipses, lines, curves and other shapes. SVG documents can be used in web pages, where they can provide interactive and animated graphics that scale without loss of quality. SVG markup can also be embedded in other XML applications, for example XSL-FO, to take advantage of the full resolution of printers, generating very accurate document printouts.

SVG was developed by a W3C working group in 1998 and SVG 1.0 became a W3C Recommendation in 2001 and SVG 1.1 in 2003. There is little difference between SVG 1.0 and 1.1, but the SVG 1.1 specification introduced a number of profiles for SVG: for example, the subsets SVG Basic and SVG Tiny (a W3C Recommendation in its own right), which are designed for use of SVG on devices of limited memory and bandwidth, such as mobile phones and personal data assistants (PDA).

The advantages of Scalable Vector Graphics

To display images in web pages, bitmapped formats such as JPEG, GIF and PNG are commonly used. However, you cannot scale bitmapped images without distortion. SVG images, on the other hand, like vector formats in general, do scale without loss of quality to fit any size of display or device. While other vector formats exist, and some, such as the proprietary SWF format of Macromedia Flash, can provide interactivity for web pages, no standard portable vector format has emerged until SVG. As an XML format, SVG provides an open vector format for the web.

SVG documents are text documents like any other XML document. As well as describing images, this text can provide metadata about the image. The textual content can easily be searched and indexed, and to help people with disabilities, it can be rendered as speech, using assistive technology such as screen readers that generate voice output. Another advantage for accessibility is that SVG can be scaled without loss of clarity, helping people with impaired vision.

SVG, like other XML applications, supports Unicode, which means that SVG documents can contain text from any of the world's languages, and multiple languages can be used in the same document.

SVG documents can include markup from other XML languages - that is, from other namespaces - and you can include SVG elements in documents written in other XML languages. For example, a Synchronized Multimedia Integration Language (SMIL) document might include SVG elements to enhance its graphical content. Also, you can include SVG documents in an XSL-FO document. In this case, the XSL-FO describes the general text-based page layout, while SVG describes the graphics.

Scalable Vector Graphics features

The SVG language supports three types of graphic object:

- *Vector graphic shapes.* The basic shape types are represented by the `<rectangle>`, `<circle>`, `<ellipse>`, `<line>`, `<polyline>`, `<polygon>` and `<path>` elements: see Figure 10.1 on page 180. The `<path>` element can be used to define the basic shapes, as well as complex shapes composed of straight lines and curves, including *Bézier curves.*
- *Raster (bitmapped) images.* Raster images can be referenced with the `<image>` element. This is useful for embedding, for example,

photographs and screen captures, which you can then overlay with text and arrows.

- *Text*. Attributes and properties on the `<text>` element indicate things such as the font specification, painting attributes and writing direction, which together describe how to render the characters. Fonts can be defined within SVG content using the `<font>` element. This guarantees the same rendering results for all users, even when the user does not have the font installed. Fonts can either be embedded within the same document that uses the font or saved as part of an external resource.

Graphic objects can be grouped and groups themselves can be nested, so that complex diagrams are built up from the basic shapes. As you will gather from the simple example in this chapter, SVG documents for complex diagrams can be very verbose.

Virtually every SVG element, *including text elements*, can be clipped, masked and transformed in various ways, and filter effects can be applied. Clipping is useful for aligning text along a path, and each object can itself act as a clipping path. Transformations include scaling, rotation and skewing, while the range of filter effects that can be applied includes blurs, lighting effects and compositing.

Styling can be specified in the `style` attribute of individual elements, or stylesheets can be embedded within the SVG content with the `<style>` element. You can apply styles using CSS stylesheets, which has the obvious advantage of repeating styling over elements, and also allows for easier maintenance.

Much of the power of SVG comes from its reuse of objects. Within the `<defs>` element in an SVG document, you can define reusable objects, for example symbols, graphic objects and groups of objects, which are instantiated later in the document with the `<use>` element. As with all XML applications, you can declare entities for text strings and external resources that are referenced in the SVG document.

Animation and interactivity

SVG graphics can be both animated and interactive, mainly because scripting and other languages have access, through the Document Object Model (DOM) application-programming interface (API), to all of the SVG elements, attributes and properties.

To make SVG animated, you can either use the `<animate>` element (which is borrowed from SMIL) or use scripting. Using the `<animate>` element may be preferable, as scripts are difficult to edit and make interchange between authoring tools harder. A useful animation feature is the ability to animate along a path, which can be very effective in flow diagrams.

To make SVG interactive, you can use server-side technologies such as the Perl scripting language and Java Server Pages (JSP), or client-side technologies, including JavaScript. An extensive set of event handlers can be assigned to any SVG graphical object, so that clicking or passing the mouse over the object can, for example, start an animation, link to another web page, change the colour of the object or perhaps cause previously hidden graphic elements to appear.

Some examples of SVG

The following figures illustrate some scalable vector graphics as they would appear in a web browser or SVG viewer, together with the corresponding source markup.

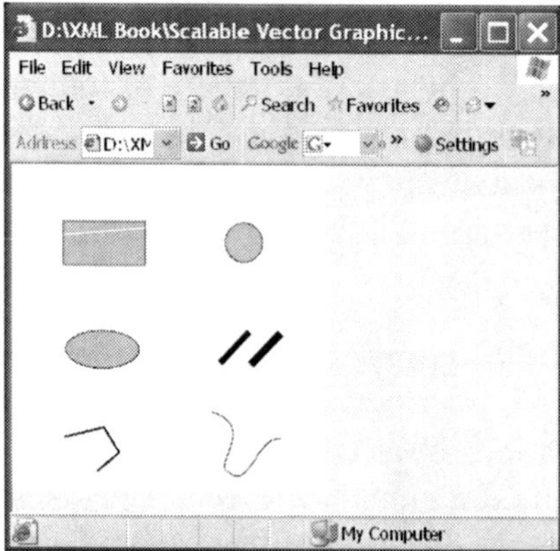

Figure 10.1 Basic shapes in SVG

Figure 10.1 shows basic shapes defined by (from left to right) the `<rect>`, `<circle>`, `<ellipse>`, `<line>`, `<polyline>` and `<path>` elements.

The following is the source markup that created the figure:

```
<?xml version="1.0"?>
<!DOCTYPE svg PUBLIC "-//W3C//DTD SVG 1.0//EN"
  "http://www.w3.org/TR/2001/REC-SVG-20010904/DTD/svg10.dtd">
<svg width="300" height="300">
  <rect x="22" y="24" width="80" height="40"
    style="fill:rgb(64,255,255);stroke:rgb(0,0,0);stroke-width:1"
    transform="matrix(1 0 0 1 88.2472 106) translate(-60.5381 -78.0269)"/>
  <circle cx="213" cy="48.5" r="18"
    style="fill:rgb(64,255,255);stroke:rgb(0,0,0);stroke-width:1"
    transform="matrix(1 0 0 1 -62.809 101.629) translate(72.6457 -76.6816)
    translate(1.34529 -2.69058)"/>
  <ellipse cx="62" cy="119.5" rx="36" ry="17.5"
    style="fill:rgb(64,255,255);stroke:rgb(0,0,0);stroke-width:1"
    transform="matrix(1 0 0 1 89 31) translate(-63.2287 18.8341)"/>
  <line x1="136" y1="165" x2="165" y2="136"
    transform="matrix(1 0 0 1 -0.0021 0.338642) translate(64.574 16.1435)"
    style="fill:none;stroke:rgb(0,0,0);stroke-width:6"/>
  <line x1="136" y1="165" x2="165" y2="136"
    transform="matrix(1 0 0 1 0.1818 0.19774) translate(94.1704 17.4888)"
    style="fill:none;stroke:rgb(0,0,0);stroke-width:8"/>
  <polyline points="51.5,184.476 90.5,175.476 105.5,198.476 83.5,215.476"
    transform="matrix(1 0 0 1 71.4373 -44.8903) translate(-72.6457 110.314)"
    style="stroke:rgb(0,0,0);stroke-width:2;fill:none"/>
  <path d="M169.5 170 C211.125 182.488 173.725 214.956 182.5 223 C206.375
    244.885 211.882 191.638 235.5 194"
    transform="matrix(1 0 0 1 -51.9756 -48.5109) translate(75.3363 104.933)"
    style="fill:none;stroke:rgb(0,0,0);stroke-width:1"/>
</svg>
```

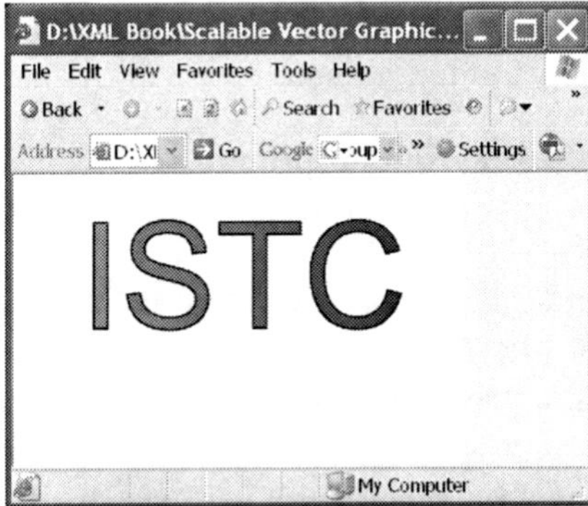

Figure 10.2 SVG Graphic showing gradient fill of text

In the following markup for Figure 10.2, note the `<defs>` element, which defines the gradient fill for the text:

```
<?xml version="1.0"?>
<!DOCTYPE svg PUBLIC "-//W3C//DTD SVG 1.0//EN"
  "http://www.w3.org/TR/2001/REC-SVG-20010904/DTD/svg10.dtd">
<svg width="200" height="200">
  <defs>
  <linearGradient id="black-white" x1="0%" y1="0%" x2="100%" y2="0%"
  spreadMethod="pad" gradientUnits="objectBoundingBox">
    <stop offset="0%" style="stop-color:rgb(0,255,0);stop-opacity:1"/>
    <stop offset="100%" style="stop-color:rgb(0,0,255);stop-opacity:1"/>
  </linearGradient>
  </defs>
  <text x="32px" y="95px" transform="translate(-2 28)"
   style="fill:url(#black-white);font-size:64;font-family:ArborText;
    fill-opacity:1;stroke:rgb(0,0,0);">ISTC</text>
</svg>
```

Producing SVG documents

You can generate SVG documents using:

- Text editors
- XML editors
- Dedicated SVG editors
- Graphics software to export SVG documents
- SVG converters
- Software such as scripting languages.

For an up-to-date list of tools that support SVG, see the SVG section of the W3C website at `www.w3.org/Graphics/SVG`.

Text editors and XML editors

SVG documents are readable text, so you can use any text editor to edit them, such as an editor like TextPad, which offers SVG syntax highlighting. However, you might prefer to use an XML editor such as XMLSpy, which provides syntax highlighting, checks for well-formedness and validates the SVG content against the SVG DTD. Text editors and XML editors are probably most useful for editing SVG documents that have already been created, which might be necessary because SVG documents exported from graphics software do not always contain optimal markup. There are other advantages to using a text editor: for example, it is easy to search and replace attribute values of elements, which could be a quick way, for example, of changing colour definitions.

There are some native SVG authoring tools, such as InkScape, which provide a WYSIWYG view in which you can draw graphics objects directly using a drawing toolbar, as well as change their properties, import raster images and so on.

There is good support for SVG in the major graphics software packages. You can import and export SVG documents with Corel Draw Version 10, Adobe Illustrator and Microsoft Visio. Adobe's web authoring tools, DreamWeaver and GoLive, also support SVG and allows the embedding of SVG into web pages, as well as editing of SVG source code. You can also import SVG documents into FrameMaker: for display, they are converted to bitmaps and the original vector format is retained for print and PDF. You can also save to SVG format from FrameMaker, or you can save SVG graphics to bitmapped formats (GIF, JPF and PNG) for web viewing.

Various tools exist for converting between SVG and various other file formats. Some of these can run as web server add-ons - for example, as *servlets* - to provide on-the-fly conversion. The SVGMaker tool from Software Mechanics acts in similar way to Adobe PDFMaker, allowing any application to print directly to an SVG document.

Software can generate SVG dynamically. For example, you can use JavaScript to generate SVG dynamically from data stored in XML documents or in JavaScript arrays. On the server side, scripting languages such as Perl and Python can be used, and Java, JSP and Active Server Pages (ASP) technologies can also generate SVG. Another means of generating SVG is to use XSLT to transform XML data extracted from databases into an SVG graphical representation of that data, for example a bar graph or a pie chart.

To view SVG documents, you need a web browser that supports SVG or a stand-alone SVG viewer. Most major browsers, including Internet Explorer, Mozilla's Firefox and Opera support at least basic SVG, either natively or though the use of an SVG plug-in such as Google Chrome Frame and the Adobe SVG Viewer. Until Version 8 at least, Internet Explorer is the only major browser that did not support SVG natively.

However, the level of support for SVG across browsers has been inconsistent, and in the early years of SVG no browser actually supported the whole SVG specification (SVG Viewer supported almost all of it, even if the SVG Viewer is not compatible with all browsers). Native support for SVG among browsers has increased, so although SVG Viewer has been the most widely used tool for viewing SVG, Adobe announced that they would no longer support the plug-in from January 2009.

The W3C's Amaya project provides an SVG editor as well as a browser. You can edit XML documents that reference multiple XML namespaces in WYSIWYG mode. The Batik SVG Toolkit from Apache provides a cross-platform, stand-alone SVG viewer, as well as tools for generating and manipulating SVG documents.

An SVG viewer is actually a rendering engine that interprets the vector description of an image and renders it as a bitmap for display on screen. For both bitmapped and vector formats, the actual display format is always a bitmap. With SVG viewers you can typically zoom in or out, scroll and pan to view the image in detail, as well as search the textual content. For SVG graphics, a different bitmap is displayed each time you resize the graphic.

Including SVG in HTML web pages

In some browsers you can include SVG directly in HTML documents, as long as the browser can render SVG. There are several methods of doing so, but in theory, you should be able to do something like the following:

```
<html>
  <head>
    <title>SVG Example</title>
  </head>
  <body>
    <h1>SVG Example</h1>
    <svg xmlns=http://www.w3.org/2000/svg style="width: 3.5in; height: 1in">
      <circle r="30" cx="34" cy="34" style="fill: red; stroke: blue;
        stroke-width: 2"/>
    </svg>
  </body>
</html>
```

However, most browsers do not support SVG markup in-line with HTML source code.

Another possibility is to link to an SVG file using an <embed> element, as in this example:

```
<html>
  <head>
    <title>SVG Example</title>
  </head>
  <body>
    <h1>ISTC SVG Graphic</h1>
    <embed src="svg example 2.svg"alt="An ISTC Logo"
      width="200" height="200" align="left"
      pluginspage="http://www.adobe.com/svg/viewer/install/"/>
  </body>
</html>
```

or an <object> element, but again both of these elements work for some browsers and not for others. The recommended method that should work in most browsers is to use both the <object> and <embed> elements:

```
<html>
  <head>
    <title>SVG Example</title>
  </head>
  <body>
    <h1>ISTC SVG Graphic</h1>
    <object data="svg example 2.svg" type="image/svg+xml"
      width="200" height="200"/>
    <embed src="svg example 2.svg" type="image/svg+xml"
      width="200" height="200" />
  </body>
</html>
```

Note that opinions vary about the best method. However, both of the latter methods work in Internet Explorer 6 and later.

The future of SVG

SVG is of great potential value in technical diagrams, where its animation and interactive features are useful in illustrating, for example, process flows and complex software relationships. SVG could also be applied in various other ways, for example in online learning services, online catalogues, multimedia and e-commerce websites. SVG might also become the standard, open format for graphics interchange, just as XML is the standard data-interchange format.

However, adoption of SVG has so far been slow: support by browsers is incomplete and inconsistent. Other technologies that support animation and other features that SVG provides are still being used. SVG will probably only become the universal web graphics format when the major browsers support SVG natively – and support the complete SVG specification.

11

XML and Content Management

This chapter discusses the advantages of XML as a storage format in content management systems (CMS). It describes:

- How the structured and semantic nature of XML allows granular content management
- How XML facilitates location and retrieval of content within a CMS
- How XML facilitates reuse of content within a CMS.

Content management

First of all, what is content management? There are many definitions to be found in the literature. *Managing Enterprise Content: A Unified Content Strategy* defines content management as:

> *The capability to manage and track the location of, and relationships among, a firm's content, at an element level in a repository.*

There are different varieties of content management systems: some allow integration with publishing tools, while others have their own publishing facilities. However, they are likely to provide the following functions (see Figure 11.1):

- Creation and authoring of content
- Saving of content with suitable *metadata*
- Storing of content in a repository
- Location and retrieval of content
- Tracking of changes to the content
- Publishing of content in different formats

- Controlling *workflow* (in high-end systems).

The advantages that XML provides for these functions are described in following sections.

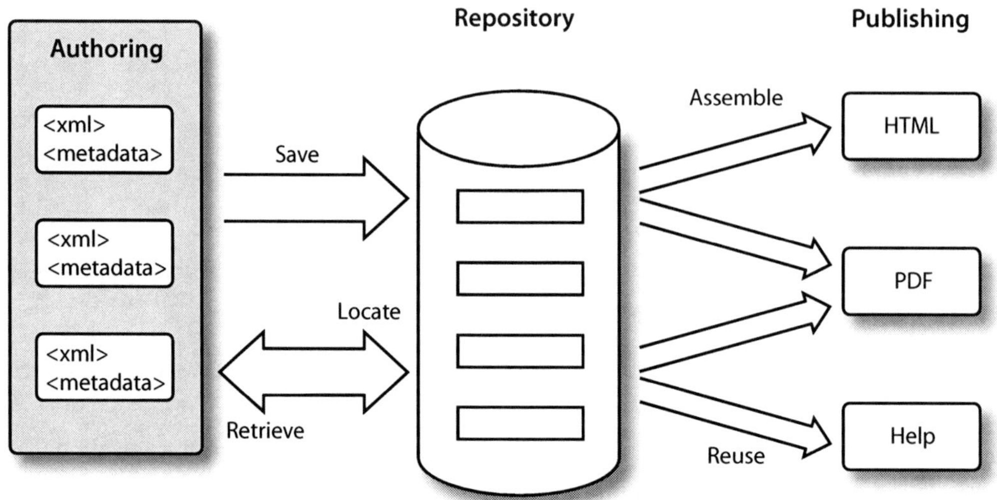

Figure 11.1 Content management functions

There are many CMS on the market, and choosing the right system for your needs is not easy. You might find the information in *Content Management Bible* and *Managing Enterprise Content: A Unified Content Strategy* useful in this respect.

Creation and authoring of content

Some CMS have their own authoring facilities, while others provide an interface so that you can check files in and out of the CMS and its repository directly from your authoring environment. If XML is not the source format, the CMS might be able to convert files to XML as they are entered into the repository.

XML is an excellent format for sharing information, because it overcomes the problem of incompatible formats and platforms. You can share content, in any natural language, between the CMS, different authoring tools and different publishing tools.

XML also has advantages for collaboration within groups of technical communicators. The fact that content can be stored as 'chunks' corresponding to topics or smaller objects, rather than whole documents, makes it easier to control access to material when people need to work on the same documents simultaneously.

Saving content with metadata

Before you save content in a CMS, it is essential to apply metadata, 'data about data', to the content. Metadata can describe not only what the content is, but how it is used and who uses it. In content management, metadata is vital for categorizing content, in retrieving and tracking content, and in allowing more effective use and reuse of the content.

Some metadata, for example the date of saving and author, is attached automatically by a CMS when the content is saved: however, XML also allows you to store metadata in its elements or attributes. You can give an element or attribute a semantic, meaningful name. For example the name of an `<audience>` element tells you what it contains information about:

```
<audience type="programmer" job="programming"
  experiencelevel="intermediate"/>
```

This example shows how attributes can further qualify what the element is used for or represents. As another example of the use of attributes for providing metadata, an `audience` attribute can provide information about the readership of a document and their experience level:

```
<p audience="beginner">Remove the DVD wrapping.</p>
```

You can use metadata to identify how components of your content are to be reused in multiple deliverables. Metadata can provide information to be used by tools and software to assemble the content for reuse and apply necessary conditional processing. The following lists some examples of such metadata, with examples of elements that might be used to provide the metadata:

- The type of content. For example: task, conceptual or reference information.
- The products to which the content applies. For example, `<prodname> DocULike</prodname>`.

- The audience for the content. For example `<audience type="administrator"/>`.
- The product release to which the content applies. For example `<vrm version="1.0" release="2007-06-30" modification="0"/>`.
- The type of information product in which the content is used, such as brochure, online help, web pages and so on. For example `<output type="PDF"/>`.

The metadata that you use to identify reusable content can also be used for categorizing and retrieving content.

Storing content

In a CMS repository, you can store XML in chunks that are smaller than a whole document, which means that you can reuse content at the level of topics, paragraphs or even smaller units. This granularity of storage allows you to extract content and reassemble it in different ways according to your needs. In some CMS the chunks can correspond to segments that are used in translation memory.

There are three basic ways in which XML can be stored in a CMS repository:

- *Object repository.* XML is stored as individual chunks.
- *Relational database repository.* XML is stored in one field of a record and the other fields of the records store metadata such as author, creation data and so on.
- *File-based systems.* XML is stored in its native format.

Relational database repositories or file-based systems provide better performance, but neither allows direct access to the XML: this must be done by retrieving the XML file and manipulating it with standard tools.

In some CMS, the metadata is extracted from XML files when they are stored in the repository. You should therefore ensure that if metadata is used in an output format such as HTML there is some mechanism for reapplying the metadata.

XML separates content from presentation, and presentation information is kept in separate files. This means that XML files are smaller in comparison to files that contain formatting information, and are therefore quicker to access.

As XML documents can include content in any natural language, some CMS allow for storage of translated content. Such CMS that are integrated with localization tools allow for more efficient management of translated material.

Locating and retrieving content

Most CMS have sophisticated tools for browsing and searching through the content they contain. You can locate the topics or other chunks of content that you require using views that organize the content and show the relationship between topics or chunks. Some CMS allow you to drag and drop objects to assemble them into topics or books.

XML allows you to store and classify information in a structured way. If you have logical groupings of content, you can find information more easily. This taxonomy of information can be enabled by metadata in the XML files that identify topic types, and perhaps information collections to which the content belongs. Also, elements and attributes in the XML can provide metadata for searching. For example, you might search by topic type, by topic title, by release state or by version. Of course XML files are text files, so you can also search them using keywords.

Tracking changes to content

Version control is an essential part of any CMS or authoring system. When you check files or objects in and out of a CMS, there must be a system of locking to ensure that authors do not overwrite each other's work. Also, files or objects must be maintained in the CMS repository with the correct version information, details of when the content was last updated and by whom. Historical information is also required so that authors can retrieve previous versions of content or view the differences between versions.

While much of the version control functionality and information is provided automatically by the CMS, XML metadata can provide information about the last modification date and perhaps details of modifications that were made. Another way in which XML is useful for showing changes to content is by using the `<rev>` element, which is used for generating revision bars in output.

Publishing content

From XML you can publish content to many different formats such as HTML, PDF and online help formats. Having information that is stored in easily located chunks means that it is easy to reuse and assemble the content required for different publications, or to repurpose the same content for different types of output format. It is the metadata associated with the XML content that allows the location of the content and defines its purpose.

Publishing to different formats is made easier by XSLT, which can apply the formatting you require and at the same time manipulate the content to re-order, repeat, filter and add information. Again, XSLT stylesheets can use metadata in their processing of XML content. For example, an XSLT stylesheet might filter XML content based on the value of attributes that identify the audience or the version of the content.

Controlling workflow

A workflow is a set of tasks that occurs in a specific sequence. There is a workflow, for example, when you create, store, edit, review and publish content such as technical documentation. Some high-end CMS can automate workflow tasks: for example, reviewers can be automatically notified by e-mail when drafts are ready for review, and content can be routed to them accordingly. As another example, when content is identified as ready for publishing, the publishing process can be started automatically.

Whether or not the workflow associated with your content management is automated, it is important to track the status of information in your information development cycle. For example, you can show that a document is ready for review, ready for editing, translated or published. By tracking content status, you facilitate collaboration within your team and with other groups.

XML metadata can be used here to provide the state information for your workflow: you might, for example, use a `<release-state>` element to indicate states such as:

- Draft
- Ready for review
- Reviewed
- Edited
- Approved
- Published.

12

Summary

This chapter summarizes:

- How XML provides facilities for handling common technical communication tasks
- The advantages that XML provides for technical communicators and how the features and facilities of XML provide those advantages.

XML and technical communication tasks

Table 12.1 summarizes how common technical communication tasks are handled and facilitated by XML. Of course, the implementation details are different for each XML language such as DocBook, DITA, or S1000D. Refer to the appropriate XML language specification for more information about the elements and attributes relevant to each task.

Table 12.1 Technical communication tasks and XML

Task	*Solution*	*References*
Producing different versions of a document for varying audiences, products, or other criteria	XML languages provide attributes to identify content as applicable for different conditions. This allows *conditional processing* in which the content for say, different products and audiences, is filtered. In this way, different versions of a document are published from a single source file.	*Conditional processing with DocBook* (page 62) *Filtering content for reuse* (page 81) *Applicability* (page 103)

Table 12.1 Technical communication tasks and XML (continued)

Task	Solution	References
Publishing to different output formats	The XML languages, XSLT, XPath, and XSL-FO and supporting software allow transformation of XML documents to different output formats such as HTML and PDF. These technologies ensure that the correct visual presentation and formatting is applied in the output.	*Publishing from DocBook* (page 64) *Tools available for working with DITA* (page 82) *The S1000D documentation process* (page 97) Chapter 7, *Transforming XML*
Managing product names and other volatile material	XML allows entities and content stored in central files to be referenced throughout a documentation set, so that product names, brand names, and other volatile content is maintained in one place, and reused in the documentation.	*Entity references (lines 9, 24)* (page 16) *Reusing elements within topics* (page 80)
Documenting a procedure	XML languages provide elements and attributes for documenting procedures. DocBook has the `<procedure>` element and its child elements, DITA has a Task topic type, and S1000D provides the procedure data module type.	*The elements of DocBook* (page 57) *The Task topic type* (page 70) *Data module content section* (page 90)
Reusing information	From reuse of words, phrases and sentences, topics (DITA), data modules (S1000D), to even whole publications, XML offers powerful facilities for reuse and repurposing of content.	*DITA and reuse* (page 78) *S1000D and reuse* (page 113)

Table 12.1 Technical communication tasks and XML (continued)

Task	*Solution*	*References*
Saving localization cost and facilitating translation	The reuse facilities of XML reduce localization costs because there is less to translate the more content is reused. Individual XML languages such as DITA provide attributes and other facilities for guiding translation and making localization easier. Many XML languages, such as XLIFF and TMX exist for facilitating the actual localization process.	*The advantages of DITA* (page 83) Chapter 9, *XML and Localization*
Including graphics	XML languages provided elements such as `<image>` for the inclusion of graphics. The Scalable Vector Graphics language is an XML language specifically for describing vector graphics.	*The elements of DocBook* (page 57) *The generic topic type* (page 67) *Common constructs in data modules* (page 92) Chapter 10, *Scalable Vector Graphics*
Identifying revisions in documentation	XML languages like DocBook and DITA provide elements and attributes (such as DITA's `rev` attribute) to show what content has changed between drafts of a document, or between published revisions of the document.	*DocBook common attributes* (page 60) *Common constructs in data modules* (page 92)

Table 12.1 Technical communication tasks and XML (continued)

Task	*Solution*	*References*
Producing an index	DocBook, DITA, and S1000D all provide elements for defining index entries, while tools such as the DITA Open Toolkit are used to generate indexes.	*The elements of DocBook* (page 57) *Tools available for working with DITA* (page 82) *Common constructs in data modules* (page 92)
Producing a glossary and defining terms	XML languages like DocBook and DITA provide elements for definitions of terms and for building glossaries from terms and their definitions.	*The elements of DocBook* (page 57) *DITA specialization* (page 77)
Including code samples and showing programming language syntax	XML languages suitable for software documentation, such as DocBook and DITA, provide a rich set of elements and specializations for documenting programming language syntax and for including code samples.	*The elements of DocBook* (page 57) *DITA domains* (page 68) *DITA specialization* (page 77)
Documenting error messages	XML languages provide elements for documenting error messages (DocBook and DITA) or fault conditions (S1000D).	*The elements of DocBook* (page 57) *The Reference topic type* (page 71) *Data module content section* (page 90)

The advantages of XML

XML's greatest strength lies perhaps in the *reuse* of content. XML has many features that facilitate reuse, both in reassembling and repurposing content, including:

- Modular structure, which makes it easier to assemble documents containing the material required.

- Entities, which allow reuse of boilerplate text and reuse of definitions within DTDs.
- Metadata built into its elements and attributes that associate content with its meaning and purpose and control conditional processing.
- Easy transformation to produce different outputs from single-source documents. XSLT and XSL-FO are powerful XML technologies for converting XML to HTML, PDF or other XML languages such as WML. XSLT can add content as it reuses information during transformation.
- Powerful reuse facilities in DITA, with its `conref` mechanism, object-oriented features and use of maps to organize content for reuse. S1000D also offers extensive reuse capabilities.

Reuse of content results in the following advantages:

- More consistent content that is easier to maintain because changes must only be made in one place
- Reduced information development cost, because there is less content to create, review and publish, and because maintenance of existing content is easier
- Increased quality, because information is more consistent, and more time can be devoted to editing and quality assurance
- Reduced translation cost for localized content: the greater the reuse of content, the less there is to translate, which lowers translation cost per language.

XML has various advantages for technical communicators:

- The separation of content from presentation, which allows you to concentrate on document content and save time on formatting (which can be a problem with, say, Word).
- A structured nature that ensures consistency in documentation. An information model can be enforced: for example, DITA encapsulates a writing model in which conceptual, task and reference topics follow a prescribed structure. The reuse of content also promotes consistency, which has a number of benefits for readers, including the ability to get used to an organization's or product's documentation, navigate it more easily and find information more quickly.
- Easy export of XML content for localization or storage. This is true even when XML is not used as the documentation source format.
- An open standard that provides a platform-independent, non-proprietary, license-free format. You can produce XML documents

using a particular program on one operating system and use the document in other programs on different operating systems. This is a great advantage for sharing material and collaborating with other groups within your organization and other organizations. DocBook, DITA and S1000D are open standards.

- The considerable advantage of understanding a technology that is so widely used in software development and data interchange. This knowledge is useful for working with XML documents used by software developers and in discussions with subject matter experts.

XML increases return on investment (ROI) for documentation projects. The ROI is the ratio of money gained or lost on the investment relative to the amount of money invested and is usually expressed as a percentage. Figures for ROI depend upon various factors including the XML solution you use, but results from cost savings in a number of areas as described in the following list:

- *Reduced development, review and maintenance cost.* Cost is reduced due to reuse of content. Less new content must be developed, meaning less time and money spent on review and maintenance. ROI is 25–80%, depending on extent of reuse.
- *Reduced translation costs.* The cost of translation is reduced by the percentage of reuse. Not only actual translation costs, but costs for reviewing of translated content and post-translation publication of documentation are reduced. The ROI increases as the number of languages increases. ROI is typically 25–50%.

The percentage figures given are examples for XML DITA suggested in *DITA 101* by Rockley *et al. DITA 101* also mentions increased ROI through greater consistency, and the ability to reconfigure information products rapidly, but gives no percentage figures for ROI from these areas.

In regulated industries and in organizations where it is necessary to comply with specific structures due to governmental or industry standards, the structure enforced by the use of XML can be vital in ensuring that your documents conform to the required standards.

By definition, XML languages are extensible: they can be extended and specialized to provide exactly the markup you require. XML languages have been developed specifically for technical communication, including DocBook and DITA: they implement structured authoring and are adaptable. You can customize DocBook to provide additional elements, (although some would argue against this) or to use only a subset of elements. In DITA, with its specialization capability, you can add markup to support new information types or new classes of markup to be used across information types. The set of XML elements available with S1000D is fixed, but a project can decide which of the many elements to use and determine the interpretation of those elements through the use of business rules. Another XML language, Scalable Vector Graphics, promises to become an open standard graphics format for use on the web and elsewhere.

XML has a number of advantages for internationalization and localization, both as a source format for content in any natural language and in the number of XML languages, such as XLIFF, TMX and ITS, that are used to support the localization process. The reuse that XML allows can lower translation cost dramatically, and translation can be made easier through the guidance that XML can provide for translators.

Due to the semantic nature of XML, the meaning inherent in element names and the metadata provided in XML documents can be exploited in support of intelligent searching of content: it may one day enable the Semantic Web, which you could search for resources according to their meaning. XML metadata is also one of the reasons why XML has a key role in content management, where it is useful in categorizing content, facilitating information retrieval and enabling easier and more effective reuse of content in any language.

Other advantages of XML include the huge support that it receives from the software industry and from standards organizations such as the W3C and OASIS, its self documenting nature and its usefulness as a storage format in various types of database.

Although there are many advantages to XML, there are some potential problems in adopting XML. Considerable cost and effort can be involved in migrating to XML. Acquiring new tools and converting existing source formats to XML formats is costly, and there can be a steep learning curve for users. Implementing publishing solutions can involve a lot of programming.

XML authoring can be daunting for authors not used to markup language: having to enter large amounts of metadata into XML documents is certainly hard work. Other shortcomings include the lack of complete and consistent support for various XML languages and technologies in web browsers, and the fact that some XML languages, such as XSLT, are not particularly easy for non-programmers to learn. However, the advantages of XML outweigh the disadvantages, and any shortcomings that it has today will be overcome.

There is no doubt that XML is here to stay. Technical communicators who understand XML will be better equipped to take advantage of all it has to offer. XML and XML-based technologies such as DITA and S1000D will play an ever-increasing role in technical communication: they will make the production of technical documentation more efficient and cost-effective, while the resulting documentation will be more consistent, more reusable and more relevant to its readers.

A

Glossary

access illustration — In *S1000D*, a data module with one or more illustrations that is used to support the navigation of an interactive publication.

applicability — In *S1000D*, a feature that allows you to specify which information in a data module is appropriate for particular conditions and product attributes, and which enables conditional processing of information to produce documentation tailored for different configurations.

application — A *markup language* that conforms to the syntax rules of a particular *metalanguage*.

For example, the markup language *DocBook* Version 3.1 is an *SGML* application, and XML Schema is an XML application.

attribute — A property of an XML *element* that provides additional data to qualify the element. Attributes are name-value pairs: in XML the value must be enclosed in single or double straight quotes.

Bézier curve — A technique used in computer graphics to produce curves that appear reasonably smooth at all scales. The most commonly used type of Bézier curve defines four points: two endpoints and two control points that do not lie on the curve itself, but define the shape of the curve. By moving the endpoints themselves, or the control points, you can modify the shape of the curve.

business rules — In *S1000D*, the decisions made for a project or by an organization on how to implement S1000D. Business rules cover not only decisions about which XML elements and attributes are used, and their interpretation, but also decisions about standards, specifications and business processes that are related to the implementation of S1000D.

Cascading Style Sheets (CSS) — A language for writing *stylesheets* that specify how HTML or XML documents should be formatted for display in web browsers.

catalog file — See *XML catalog file.*

CDATA section — An XML markup item used to identify data that processing software should not interpret as markup. CDATA sections are used to include sections of XML, HTML and other code within XML documents. A CDATA section begins with `<![CDATA[` and ends with `]]>`.

character reference — An XML markup item used to insert single characters such as special characters and symbols that you cannot otherwise enter into an XML document. A character reference begins with an ampersand (`&`), followed by a hash sign (`#`) and a number to specify the character, and ends with a semicolon (`;`).

For example, the character reference `Ç` is used for the capital C character with a cedilla (Ç).

child element — An *element* contained within another element.

Common source database (CSDB) — In *S1000D*, the repository for storing and managing the data modules and other information associated with documentation projects.

complex type — In the *XML Schema language*, a type in which the *element* can contain child elements and *attributes.*

content management — The capability to manage and track the location of, and relationships among, a firm's content, at an *element* level in a repository. (From *Managing Enterprise Content: A Unified Content Strategy.*)

content specification — A definition of what an XML *element* can contain in terms of character data and *child elements.* Also known as a *content model.*

CSS — See *Cascading Style Sheets.*

customization layer — In some versions of *DocBook*, a DTD that references some or all of the DocBook modules. The customization layer allows you to modify the DTD by adding your own declarations that override declarations in the DTD modules.

Darwin Information Typing Architecture (DITA)	An XML framework for the production of *topic-oriented* technical documentation. DITA is not only an *XML language*, but also an architecture for designing information that encapsulates best practices and extensible design.
data module	In *S1000D*, a self-contained unit of information that consists of an identification and status section and a contents section. The identification and status section contains metadata used in managing the data module and a data module code used in identification. The contents section contains the information seen by the user, information supporting data management and reuse, or, in one type of data module, the business rules for managing the documentation project. Data modules are central to reuse in S1000D documentation projects.
derived type	In the *XML Schema language*, a type based on one of the built-in *simple types*.
DITA map	A file that contains a set of references to *DITA* topics, and which allows you to organize the topics into a hierarchy and define the relationships between the topics.
DITA Open Toolkit	An open source implementation of the *OASIS* specification for *DITA*. The DITA Open Toolkit is a collection of sample files, *stylesheets* and tools for working with DITA.
DocBook	A *markup language* for technical documentation. The original version of DocBook was an *SGML application*, and later versions were developed as XML applications.
document element	Synonymous with *root element*.
document object model (DOM)	An application programming interface (API) that represents an XML document as a tree of nodes. The DOM is used by *XML parsers*.
Document Style Semantics and Specification Language (DSSSL)	A standard *stylesheet* language for *SGML* and XML documents (ISO/IEC 10179:1996).
document type	Synonymous with *XML language* or *XML vocabulary*.
document type declaration	In an XML document, the statements that declare the *document type* – or in other words, the *XML language* in which the document is written.

Document Type Definition (DTD) — A file that defines the *elements*, *attributes* and other markup that can appear in an XML document of a particular type. As for other types of XML *schema document*, a DTD specifies the type and structure of information that can appear in a *valid* XML document that conforms to the DTD. 'DTD' refers both to the actual files and the language used to write the DTD files.

domain — In *DITA*, a set of *elements* associated with a particular subject area or authoring requirement.

DSSSL — See *Document Style Semantics and Specification Language.*

DTD — See *Document Type Definition.*

element — The XML markup item that defines the hierarchical structure of an XML document. Most elements consist of a *start tag* and an *end tag* that enclose the content of the element. *Empty elements* have only a start tag and do not have any content, although they can have *attributes.*

element content — The material in an XML *element* between the start and *end tags.*

Element Definition Document (EDD) — In FrameMaker, the document that defines the *elements* that are allowed in a structured FrameMaker document. An EDD is the equivalent of an XML DTD.

empty element — An XML *element* that contains neither character data nor *child elements.* An empty element may have *attributes.*

end tag — The *tag* at the end of an XML *element.* An end tag has the syntax `</name>`.

entity — In a DTD, a name for data or markup that can be referenced. There are two main types of entity: *general entities*, which are declared in a DTD and referenced in XML documents, and *parameter entities*, which are declared and referenced within a DTD.

A *general entity* can represent a single character, a string of text, an XML file or a graphics file. Parameter entities can represent *element* and *attribute* declarations or whole DTDs. Entities are used for content reuse and to minimize document size.

See *general entity*, *parameter entity*, *internal entity*, *external entity*, *parsed entity* and *unparsed entity.*

entity reference	A reference to an *entity* defined in a DTD. References to general entities are made in XML documents and have the `&entity;` syntax, while references to parameter entities are made within a DTD and have the `%entity;` syntax. When an XML document is processed, entity references are replaced by the value from the corresponding entity definition in the DTD.
enumeration	A *type* of *attribute* that specifies that the attribute can contain a value from a list of possible values (separated by the '`|`' character).
Extensible HyperText Markup Language (XHTML)	A standard extensible *markup language* for web pages that is similar to HTML but is based on XML rather than *SGML*.
Extensible Stylesheet Language (XSL)	The *XML language* used to transform XML documents into a suitable format for display. XSL has two parts: XSL Transformations (*XSLT*) and *XSL Formatting Objects* (XSL-FO).
external entity	An *entity* in which the entity content is contained in a separate file from the DTD. External entities are used, for example, to include boilerplate text in XML documents, or to include chapters in books.
external subset	The part of a DTD that is contained in a separate file rather than within the *document type declaration* of an XML document.
flow	In the *XSL-FO* language, a portion of content to be laid out in a region of a page.
FO processor	Software that converts an *XSL-FO* document into a printable or readable format. This is usually PostScript or PDF, but formats such as RTF or other print-ready forms can also be produced.
frequency	The number of times *child elements* can appear within the *parent element*, as specified using one of the *occurrence indicator* characters: question mark (?), asterisk (*) and plus sign (+).
general entity	An *entity* that is declared in a DTD and referenced in XML documents. Compare with *parameter entity*.
Global Information Management Metrics Exchange (GMX)	An XML standard for word and character counts and for the exchange of localization metrics.

instance document	An XML document based on the *XML Schema language.*
Interactive Electronic Technical Publication (IETP)	A set of information needed for the description, operation and maintenance of a product, optimally arranged and formatted for interactive screen presentation to the user. IETP includes conditional branching mechanisms, which can be based on user feedback.
internal entity	An *entity* in which the entity content is contained in the DTD.
internal subset	The part of a DTD that is contained within the *document type declaration* of an XML document.
Internationalization Tag Set (ITS)	An *XML language* and *W3C Recommendation* developed by the W3C Internationalization ITS Working Group that addresses the need for *localization* directives in XML documents.
LISA	See *Localization Industry Standards Association.*
localization	The translation and adaptation of a product to different cultural conventions. Localization involves not only the translation of technical publications from English to a number of other languages, but also the translation of all localizable material in software products, such as user interface strings, error messages and other text.
Localization Industry Standards Association (LISA)	An organization responsible for developing *localization* industry standards, including *XML languages* used in localization such as *TMX, TBX* and *SRX.*
markup language	A set of codes or tags that surrounds content and describes what that content is, or in some cases what it should look like when displayed.
metadata	Data about data. The *attributes* of *elements* can constitute metadata by providing further information about the content of the element.
metalanguage	A language for defining *markup languages.* XML and *SGML* are both metalanguages.
mixed content	A *content specification* for an *element* that contains a mixture of character data and elements.
namespace	See *XML namespaces.*

notation	A declaration in a DTD that provides information about handling unparsed entities. A notation can include helper information that software can use to process the *entity's* content.
occurrence indicator	A character in the *content specification* for an XML *element* that defines how many times a *child element* can appear within a *parent element*. The occurrence indicators are question mark (?), asterisk (*) and plus sign (+).
ontology	A formal description of resources and their relationships.
Open Architecture for XML Authoring and Localization (OAXAL)	An initiative of OASIS to encourage the development of an open-standards approach to XML authoring and localization.
Open Standards for Container/Content Allowing Reuse (OSCAR)	A working committee of *LISA* that has been responsible for developing many of the *XML languages* used in *localization*.
Organization for the Advancement of Structured Information Standards (OASIS)	A not-for-profit consortium that fosters the development and adoption of open standards such as *DocBook*, and more recently, *DITA*.
parameter entity	An *entity* that is declared and also referenced in a DTD or set of DTDs. Parameter entities are used to customize DTDs and to control marked sections of a document. Compare with *general entity*.
parent element	An XML *element* that contains other elements, which are known as *child elements*.
parsed character data	XML *element content*, which includes simple text and can include character and *entity references*.
parsed entity	An *entity* in which the entity content is in XML format and can be parsed by processing software.

processing instructions
An XML markup item used to pass information to software that processes the XML document. For example, a processing instruction can specify the name of an *XSLT stylesheet* to be used by an *XSLT processor.*

publication module
In *S1000D*, an XML file that organizes the data modules that are required for a publication. A publication module can reference data modules, other publication modules and legacy publications.

read/write rules file
In FrameMaker, a file for describing how *elements* and *attributes* are translated during import and export of *markup language* documents.

reassembly
The use of modules of content in a number of documents. In reassembly, separate modules are maintained in one place and assembled into different documents as and when required, possibly at the time of publication.

relationship table
In *DITA maps*, a section of markup that defines the relationship between topics.

RELAX NG
An XML language for writing schema documents that also offers a compact, non-XML syntax. RELAX NG stands for REgular LAnguage for XML Next Generation.

repurposing
The delivery of documentation in multiple output formats from a single source format. For example, you might use the same content in online help, a PDF document and in web pages.

root element
The first *element* in an XML document that contains all the other elements in the document. Also called the *document element.*

round tripping
The export and import of XML files from and to a software application. Round tripping can refer to exporting and importing XML in FrameMaker and to the exporting and importing of XML for translation from a particular authoring tool.

S1000D
An international specification for the production and procurement of technical publications. S1000D data modules can be coded in *SGML* or XML, for which DTDs or XML Schemas are provided, depending on the issue of the S1000D specification.

SAX
See *Simple API for XML.*

Scalable Vector Graphics (SVG) — An *XML language* for describing two-dimensional vector graphics.

schema document — A description of the markup allowed for an *XML language.* DTD and *XML Schema Language* are the most common languages for writing schema documents and therefore defining XML languages. By validating XML documents against a schema document, you can ensure that all XML documents of a particular language have a consistent structure.

schema language — A language for writing *schema documents*, which define the *elements*, *attributes* and other items that can appear in a document marked up in a particular *XML language.* The two main schema languages in use are *Document Type Definition* (DTD) and the W3C *XML Schema language.*

Segmentation Rules eXchange (SRX) — An *XML language* used for defining how *translation memory* tools segment source texts to produce their databases.

semantic tagging — The feature of XML documents that allows for a hierarchical structure of *elements* with meaningful names.

Semantic Web — A vision of a World Wide Web in which you can search for information and resources according to their meaning.

servlet — A small program that runs within a web server that receives and responds to requests from web clients.

SGML — See *Standard Generalized Markup Language.*

Shareable Content Object Reference Model (SCORM) — A collection of standards and specifications for Web-based e-learning. SCORM is a specification of the Advanced Distributed Learning (ADL) Initiative of the Office of the United States Secretary of Defense.

Simple API for XML (SAX) — A standard application programming interface (API) used by *XML parsers.*

simple type — In the *XML Schema language*, a type in which the *element* can contain only text and cannot have *attributes.*

single source publishing — A publishing methodology that allows the publishing of documentation to multiple output formats from a single source format. XML has a number of features that make it an ideal source format for single source publishing. Also known as *single-sourcing.*

smart document — A Microsoft Word document that is programmatically customized with built-in intelligence to help users create and update documents.

specialization — In *DITA*, the creation of new *elements*. There are two types of specialization in the DITA specification: structural specialization, which is used to develop new information types, and *domain* specialization, which is used to develop domain-specific markup.

SRX — See *Segmentation Rules eXchange*.

Standard Generalized Markup Language — A *markup language* developed in the 1960s to provide a standard format for the sharing of documents between different programs.

start tag — The *tag* at the start of an XML *element*. A start tag has the syntax `<name>` and may optionally include *attributes*.

structured application — A document that specifies the location of the files that control how FrameMaker documents are transformed into *markup language* and vice versa. Also known as a *structure application*.

structured document — In FrameMaker, a document that uses a defined structure based on an *Element Definition Document* (EDD), in which formatting is handled automatically according to the EDD.

stylesheet — A file that specifies how an XML document should be presented for display on the web or on the printed page. Several languages for writing stylesheets are available, including CSS, *DSSSL* and *XSL*.

SVG — See *Scalable Vector Graphics*.

tag — An *element* name enclosed by angle brackets. In XML, an element has a *start tag* with the syntax `<name>` and an *end tag* with the syntax `</name>`, unless the element is an *empty element*, in which case it has only a start tag with the syntax `<name/>`.

tag minimization — In *SGML*, the facility by which DTD declarations can include the *tag* omission characters (– and o) to indicate whether or not start and *end tags* are required in an *element*. The characters are the hyphen and the letter 'o'.

template rules — In an *XSLT stylesheet*, a section of markup that uses an *XPath* expression to specify a pattern of nodes and a template to be output when the pattern is matched. The template can include markup, content from the input document and new content.

TermBase eXchange (TBX) — A standard *XML language* for the interchange of terminological data with terminology databases.

topic-oriented — Pertaining to documentation that is organized as collections of discrete topics, as opposed to a single, continuous narrative such as a book.

translation memory — A repository of text that has been translated.

Translation Memory eXchange (TMX) — A standard *XML language* for the exchange of *translation memory* data created by *localization* tools.

type — In the *XML Schema language*, the type of data that an *element* contains. The type can be a *simple type*, in which case the element can contain only text and cannot have *attributes*, or a *complex type*, in which the element can contain *child elements* and attributes. A type in XML Schema is equivalent to the *content specification* of an element in a DTD.

Unicode — A character encoding system that embraces most of the world's languages. As XML supports Unicode, XML documents can contain characters from all the natural languages covered by Unicode.

Unicode Transformation Format (UTF) — An encoding scheme for *Unicode* that specifies the digital representation of characters in the Unicode character set. UTF-8 and UTF-16 are two variants used in XML documents. UTF-8 is a superset of ASCII and is the default encoding for XML documents.

unparsed entity — An *entity* in which the entity content is not in XML format and cannot be parsed by processing software: for example, graphics files.

unstructured document — In FrameMaker, a document that does not use an *EDD*-based structure and uses paragraph and character tags to control formatting.

valid — For XML documents, a document that conforms to the rules defined in a DTD or other *schema document*.

validating editor	An XML editor that validates an XML document as it is being authored. A validating editor shows which *elements* and *attributes* are *valid* at the current cursor position.
validating parser	An *XML parser* that checks whether the parsed XML document is *valid* in addition to being *well formed.*
well formed	For XML documents, a document that conforms to the rules of XML syntax.
W3C Recommendation	The final stage in the process of ratification of a standard developed by the *World Wide Web Consortium.* A Recommendation is equivalent to a published standard in other industries.
Wireless Markup Language (WML)	An *XML language* for storing information for display on mobile devices.
workflow	A set of tasks that occurs in a specific sequence.
World Wide Web Consortium (W3C)	A consortium of member organizations and individuals that is responsible for the development of XML, key *XML languages* and supporting technologies, as well as standards for the World Wide Web.
XHTML	See *Extensible HyperText Markup Language.*
XLIFF	See *XML Localization Interchange File Format.*
XLink	An *XML language* for defining links between and within XML documents and other documents.
XML catalog file	A file that maps `PUBLIC` identifiers for DTDs to a specific location on a given machine. If you move the DTD to a new machine, it is only necessary to edit the *catalog file*'s mapping of the `PUBLIC` identifier rather than edit all of the XML documents that reference the DTD.
XML declaration	The statement in an XML document that specifies the version of XML and the character encoding used in the document.
XML language	A *markup language* that uses XML syntax. Examples of XML languages include *XSLT*, *XML Schema Language* and *SVG.* XML languages are also called XML *vocabularies* or XML *applications.*

XML Localization Interchange File Format (XLIFF) — An *XML language* that provides a common format for exchanging *localization* data between software organizations and localization vendors, or between localization tools such as *translation memory* systems and machine translation systems.

XML name — A syntax for the names of XML *elements* and *attributes*. An XML name can contain any alphanumeric characters, both English characters and characters from other languages, as well as the underscore (_), hyphen (-) or period (.) characters.

XML namespaces — A mechanism for allowing the use of *elements* and *attributes* of the same name from multiple vocabularies in a single XML document.

XML parser — Software that analyses and processes XML documents. A parser reads an XML document and determines the structure and properties of the data. An XML parser always checks that the XML document is *well formed* and may also validate the document against a DTD, in which case it is termed a *validating parser*.

XML Schema language — An *XML language* for writing *schema documents*. Like other types of schema document, an XML Schema document specifies the type and structure of information that can appear in a *valid* XML document that conforms to the schema.

XML vocabulary — Synonymous with *XML language*.

XPath — A language for locating parts of an XML document, used by *XSLT* and other *XML languages*.

XSL — See *Extensible Stylesheet Language*.

XSL Transformations (XSLT) — An *XML language* that specifies rules for transforming XML into another format, the rules being supplied in an XSLT *stylesheet* that is associated with an XML document. XSLT can transform XML documents to HTML, other XML languages, or any text-based documentation format.

XSLT processor — Software that processes an XML document and transforms it using an *XSLT stylesheet*.

XSL Formatting Objects (XSL-FO) — An *XML language* for describing the layout of text on a page (screen page or paper page). XSL-FO documents are usually produced by transformation from XML documents and are used for further transformation to formats such as PDF.

B

Bibliography

This appendix lists key XML-related web resources, books and articles.

Web resources

General XML

`www.apache.org` The home page of the Apache open source software project, which produces software such as XML parsers.

`xml.coverpages.org` The Cover Pages website, hosted by OASIS, which provides online resources for XML standards and technologies.

`www.oasis-open.org` The home page of the OASIS open standards consortium. OASIS fosters the development of both DocBook and DITA.

`www.xml.com` The O'Reilly and Associates website containing links to XML-related articles, books and blogs.

`www.w3.org` The home page of the World Wide Web Consortium (W3C).

`www.w3.org/TR/REC-xml` The XML 1.0 Recommendation on the W3C website.

Scalable Vector Graphics

`www.adobe.com/svg` Adobe's SVG web pages containing information of general SVG interest and information about the Adobe SVG Viewer, which you can download from this site.

`xml.apache.org/batik/svgviewer.html` The web page for the Batik SVG browser (also known as Squiggle).

`www.svgmaker.com` The website for downloading SVGMaker, a tool that allows applications to print to SVG documents.

`www.w3.org/Amaya/Amaya.html` The W3C Amaya project website, which provides software for browsing and creating SVG documents.

`www.w3.org/Graphics/SVG` The W3C SVG home page, containing the SVG specifications and links to SVG tools and other resources.

`www.xsmiles.org` The home page of X-Smiles, a browser that supports both SVG and the Synchronised Multimedia Integration Language (SMIL).

XML certification

`www-03.ibm.com/certify/certs/xm_index.shtml` IBM's website for XML certification.

`www.xmlmaster.org/en/` A website for an XML certification program that provides two levels of certification through examination.

XML documentation languages and specifications

`www.ditausers.org` A user group website that provides a newsletter, tutorials, a wiki and other resources.

`www.docbook.org` The website from which you can download the DocBook schemas and documentation.

`docbook.sourceforge.net` A website containing DocBook resources such as XSL stylesheets.

`wiki.docbook.org/topic` A Wiki devoted to DocBook.

`www.jclark.com` James Clark's home page, from which you can download the Jade DSSSL tool, the XP XML parser and other DocBook resources.

`www.oasis-open.org/docbook` The OASIS DocBook home page, containing a repository of DTDs and schemas for all versions of DocBook.

`dita.xml.org` The website and online community for the DITA OASIS standard.

`dita.xml.org/standard` The web pages from which you can download the DITA Language Specification and Architectural Specification documentation.

dita-ot.sourceforge.net The website from which you can download the DITA Open Toolkit, which is used to transform DITA content into a variety of formats.

dita-ot.sourceforge.net/SourceForgeFiles/doc/user_guide.html The web page for downloading the DITA Open Toolkit User Guide.

www.ditaworld.com A website containing links to many DITA resources, including articles, specifications, events, vendors and tools.

www.s1000d.org Official resource for S1000D information.

www.s1000d.net S1000D information resource site.

www.s1000dworld.com S1000D user site.

www.s1000dexpert.com S1000D practitioner site.

XML localization

www.w3.org/TR/2007/WD-xml-i18n-bp-20071031 The W3C web pages describing best practices for XML internationalization.

www.w3.org/TR/its The W3C website containing the Internationalization Tag Set (ITS) specification.

www.lisa.org The Localization Industry Standards Association (LISA) home page. The site includes information about open standards such as TBX, TMX and SRX.

XML migration

www.xml.com/pub/a/98/07/dtd/index.html A web page describing conversion of SGML DTDs to XML.

tidy.sourceforge.net The website from which you can download HTML tidy, a tool that helps in creating XHTML from HTML.

xml.lander.ca/sgml_xml_cs/sgml_xml_cs.html A web page describing SGML to XML migration strategies.

www.w3.org/TR/NOTE-sgml-xml.html The W3C web page comparing XML and SGML.

XML tools and transformations

`projects.apache.org/projects/fop.html` The website for the Apache Formatting Objects Processor. This FO processor converts XSL-FO documents into printable or readable formats.

`xerces.apache.org/xerces2-j/` The website for downloading the Apache Xerces-J parser (formerly the IBM XML4J tool).

`dsssl.netfolder.com` A website containing DSSSL resources including articles, scripts and other tools.

`tidy.sourceforge.net` The web page from which you can download HTML tidy, a tool that helps in creating XHTML from HTML.

`www.jclark.com` James Clark's home page, from which you can download the Jade DSSSL tool, the XP XML parser and other DocBook resources.

`www.thaiopensource.com/relaxng/trang.html` The web page from which you can download the Trang schema conversion tool, which allows you to convert DTDs to XML Schema documents.

XML and the web

`www.webdevout.net` A website containing detailed information about browser support of CSS and XHTML.

Books and articles

General XML

Harold, Elliotte Rusty and Means, W. Scott. *XML In A Nutshell*, O'Reilly, 2004

Van der Vlist, Eric. *XML Schema*, O'Reilly, 2002

Scalable Vector Graphics

Eisenberg, David. *SVG Essentials*, O'Reilly, 2002

Adobe FrameMaker publications available in the online manuals and XML cookbook installation folders (titles can vary with the version of FrameMaker):

Structure Application Developer Guide

Using the DocBook Starter Kit Online Manual

The Adobe FrameMaker XML Cookbook

Ethier, Kay. *XML and FrameMaker*, Apress, 2004

Lenz, Evan, McRae, Mary and St. Laurent, Simon. *Office 2003 XML*, O'Reilly, 2004

Boiko, Bob. *Content Management Bible*, Wiley, 2005

Rockley, Ann. *Managing Enterprise Content: A Unified Content Strategy*, New Riders Press, 2002

Stayton, Bob. *DocBook XSL: The Complete Guide*, Sagehill Enterprises, 2007.
Also available online at `www.sagehill.net/docbookxsl`.

Walsh, Norman and Muellner, Leonard. *DocBook: The Definitive Guide*, O'Reilly, 1999.
This version of the book only covers up to Version 3.1.5, but online revisions to the book are available on the `DocBook.org` website.

Walsh, Norman and Hamilton, Richard L. *DocBook 5: The Definitive Guide*, O'Reilly, 2010.

Larner, Ian. *Touching DITA* articles, ISTC *Communicator*.
Available, for members only, from the ISTC website at `www.istc.org.uk`

Rockley, Ann, Cooper, Charles and Manning, Steve. *DITA 101*, lulu.com, 2009.

Vazquez, Julio. *Practical DITA*, lulu.com, 2009.

XML localization

Savourel, Yves. *XML Internationalization and Localization*, Sams, 2001

XML migration

Adobe publication: *Migrating from Unstructured to Structured FrameMaker*

Available from `http://www.adobe.com/devnet/framemaker/pdfs/migrationguide.pdf`

XML transformations

Pawson, Dave. *XSL-FO*, O'Reilly, 2002

Tidwell, Doug. *XSLT*, O'Reilly, 2008

Index

A
Adobe SVG Viewer 183
animation, with SVG 179
Ant tool 83
applicability cross-reference table (ACT) 103, 107
applications, SGML 1
Arbortext 7, 83, 126, 136
attribute 14

B
BREX data module, S1000D 100
business rule, definition 98
business rules data module, S1000D 100

C
Cascading Style Sheets 161–163
  browser support 163
  example 162
  localization usage 168
  syntax 162
  versions 162
  XML documents, associating with 162
  XSLT, using with 162
catalog file 65, 66, 72
CDATA sections 18
change marking 60, 93
character references 16
child elements 13
comments, in XML documents 18
common source database, definition 112
complex types 43
conditional cross-reference table (CCT) 103, 108
conditional processing 14, 60, 63, 81
container data module 115
content management
  authoring content 188
  controlling workflow 192
  creating content 188
  definition 187
  features, content management systems 187
  locating content 191
  metadata 189
  publishing content 192
  retrieving content 191
  saving content 189
  searching content 191
  storing content 190
  tracking changes 191
  translated content 190
  XML advantages for 187
  XSLT stylesheets 192
content management systems
  choosing 188
  functions of 187
  repositories 190
cross-references 15
cross-referencing 59, 68, 93
CSDB status list 102
customization layer, DocBook 62

D
data dispatch note (DDN) 113
data module code 95
data module content section 90
data module identification and status section 87

data module requirements list (DMRL) 102
data module, common constructs 92
data module, S1000D 86
declarations, XML 13
DITA 66-85
  advantages 83
  attributes 69
  `collection-type` attribute 74
  compared with DocBook 84
  compared with S1000D 118
  Concept topic type 69
  conditional processing 81
  `conref` attribute 80
  DITA maps 72
  domains 68
  DTDs 72
  filtering content 81, 83
  FrameMaker support 83, 130
  Generic topic type 67
  maps 72
  overview 66
  Reference topic type 71
  relationship tables 75
  reuse 78
  schemas 72
  specialization 67, 77
  Task topic type 70
  tools 83
  topics 67
DITA Open Toolkit 83
DocBook 55-66
  advantages 65
  attributes 60
  compared with DITA 84
  conditional processing 63
  customizing 62
  DocBook Project 64
  DTD 61
  elements 57
  FrameMaker support 64, 130
  markup 56
  modules 61
  publishing 64
  Simplified DocBook 57
  tools 64
  usage 56
document elements 13
document interchange 172
Document Object Model (DOM) 53
Document Style Semantics and Specification Language (DSSSL) 64, 141
document type declarations 13
document type definitions (DTD)
  `ANY` element declarations 25
  attribute declarations 25
  attribute defaults 28
  `CDATA` attributes 26
  child-only element declarations 23
  comments 34
  conditional sections 35
  element declarations 22
  empty element declarations 23
  `ENTITY` attributes 27
  entity declarations 29
  enumeration attributes 27
  example 37
  external entities 29
  frequency 24
  general entities 29
  ID attributes 26
  `IDREF` attributes 26
  `IDREFS` attributes 26
  internal entities 29
  mixed content element declarations 24
  `NMTOKEN` attributes 28
  `NOTATION` attributes 27
  notation declarations 33
  occurrence indicators 24
  overview 22
  parameter entities 29
  `PCDATA` elements 23
  XML documents, associating with 36
document types 21
DOM 53
DSSL, see *Document Style Semantics and Specification Language*
DTD, see *document type definitions*
Dublin Core 87

EDD, see *element definition document*
editors, XML
  Arbortext 7, 83, 126, 136
  Authentic 128
  MadCap Flare 126
  XMetal 6, 83, 126, 141
  XMLSpy 6, 126, 144, 182
element definition document (EDD) 129, 142
empty elements 13, 23, 45

end tags 13
entities 29
entity references 16
error messages, documenting 59, 67, 71
expansion packs 135

filtering content, in DITA 81, 83
Flash, Macromedia 178
FO processors 154
formatting objects 154
FrameMaker
  API client 131
  conversion tables 142
  DITA support 131, 143
  DocBook support 131, 143
  element definition document (EDD) 129, 142
  exporting from unstructured FrameMaker 130
  exporting to XML 129
  importing from XML 129
  migrating from unstructured to structured environment 142
  migrating to XML 141
  preparation for migration 142
  problems with exporting to XML 133
  read-write rules 131, 143
  round tripping 129
  S1000D support 131, 143
  scripting languages 143
  structured applications 130, 143
  structured environment 129
  SVG support 183
  tools for XML migration 144
  unstructured environment 129
  XML output, customizing 132
  XSLT stylesheets 131
frequency, child elements 24

Global Information Management Metrics Exchange (GMX) 175, 205
glossary, elements for 58, 77
graphics
  DITA elements 68
  DocBook elements 59
  S1000D elements 94
  Scalable Vector Graphics 177

HTML
  compared with XHTML 159
  HTML Tidy tool 139
  migrating to XML 139

indexing 57, 83
information sets, S1000D 102
information types 67
instance documents 39
Interactive Electronic Technical Publications (IETP) 98
interactivity, with SVG 179
Internationalization Tag Set (ITS) 170

Jade tool 64
James Clark, tools 216, 218

List of Applicable Publications (LOAP) 110
localization
  best practices for XML internationalization 176
  Cascading Style Sheets 168
  Internationalization Tag Set 170
  reuse of content 169
  translation memory 174
  translation workflow 167
  XML character encoding 166
  XML, advantages of 165
  XML, role of 172
  XSLT stylesheets 168
Localization Industry Standards Association (LISA) 172

Macromedia Flash 178
MadCap Flare 126
Mathematical Markup Language (MathML) 4
metadata
  advantages 4
  content management, role in 189
  controlling workflow 192
  reuse of content 189
metalanguages 1

Microsoft Word
DITA support 134
DocBook support 134
expansion packs 135
migrating to XML 144
smart documents 135
tools for XML migration 144
WordProcessingML (WordML) 133
XML editor 134
XML Schema language support 134
migrating to XML
Document Style Semantics and Specification Language (DSSSL) 141
from FrameMaker 141
from HTML 139
from SGML 140
from Word 144
from XML languages 145
general considerations 137
HTML Tidy tool 139
in-house conversion 138
legacy content, cleaning up 139
outsourcing 138
strategy 138
structured writing considerations 138
XML language, choosing 137
mixed content 13, 24
MSMXL parser 53

N

name tokens 28
namespaces 20, 39, 40, 48, 49, 50, 53, 134, 150, 159, 178, 184
notations 27, 49

O

OASIS standards 172
DITA 5, 66
DocBook 5
occurrence indicators 24
ontologies 5
Open Architecture for XML Authoring and Localization (OAXAL) 176, 207
Open Standards for Container/Content Allowing Reuse (OSCAR) 172
Organization for the Advancement of Structured Information Standards (OASIS) 6, 55

P

parent elements 13
parsed character data 23
parsers, XML 52
procedures, documenting 59, 70, 91
processing instructions 17
processors, XSLT 151
product cross-reference table (PCT) 103, 109
product names, managing 16, 80
publication module code 112

R

Really Simple Syndication (RSS) 4
reassembly 9
RELAX NG 21
repositories, in CMS systems 190
repurposing 9
Resource Description Framework (RDF) 5, 87
return on investment (ROI) 198
reuse
advantages of 197
`conref` attribute, DITA 80
DITA maps 81
entities 196
filtering content, DITA 81, 83
localization reuse 169
mechanisms 193
metadata 196
reassembly 8
repurposing 8, 14
S1000D data modules 113
Scalable Vector Graphics 179
revision marking 60, 93
root elements 13
round tripping 129
RSS 4

S

S1000D 85, 85–124
advantages 118
applicability 103
applicability branches 104
applicability cross-reference table (ACT) 103, 107
BREX data module 100
business rules 98
compared with DITA 118
conditional cross-reference table (CCT) 103, 108

container data module 115
CSDB status list 102
data dispatch note (DDN) 113
data module 86
data module code (DMC) 95
data module requirements list (DMRL) 102
data module types 90
documentation process 97
FrameMaker support 117, 130
global applicability 106
information sets 102
inline applicability 106
List of Applicable Publications (LOAP) 110
overview 85
product cross-reference table (PCT) 103, 109
publication module 110
publication module code 112
reuse 113
schemas 86
SCORM content package 112
specification 86
standard numbering system 96
tools 117

Scalable Vector Graphics 177–186
adoption 185
advantages 178
animation 179
browser support 183
converters 182
editors 182
examples 180
features 178
graphics objects 178
HTML pages 184
interactivity 179
producing 182
reuse 179
support for FrameMaker 183
SVG Basic 177
SVG Tiny 177
tools 182
viewing 183
XSL-FO markup 178
XSLT 183

schema documents 4, 21
Schema languages 21
SCORM content package 112
Segmentation Rules eXchange (SRX) 175
segments, translation 174
semantic tagging 4
Semantic Web 5, 199
SGML 1, 55
migrating from 140
migrating to XML 140
table model 61
XML DTDs, migrating to 141

SGML applications
DocBook 1
HTML 2
IBMIDDOC 1

Shareable Content Object Reference Model (SCORM) 96, 112
Simple API for XML (SAX) 53
simple types 41
Simplified DocBook 57
single-source publishing 2, 8
smart documents 135
SMIL 178
specialization, DITA 67, 77
SRX 175
Standard Generalized Markup Language, see *SGML*
standard numbering system, S1000D 96
start tags 13
structured application file, `structapp.fm` 130
structured application, FrameMaker 130
SVG, see *Scalable Vector Graphics*
Synchronized Multimedia Integration Language (SMIL) 178

T

technical communication tasks 193
TermBase Exchange (TBX) 175
tools
DITA Open Toolkit 66, 83
DocBook tools 64
FrameMaker conversion tools 144
HTML tidy 139
Jade 64
S1000D 117
text editors 125
Trang 134
Word conversion tools 145
Xerces 52, 64
XML editors 126
XML4J 64
XML-aware tools 125
XP 64

topic map 77
topic types, DITA 67
Trang tool 134
translation memory 174
Translation Memory eXchange (TMX) 174

Unicode character encoding system 13, 166
Unicode Transformation Format (UTF) 167

validity 18
version control 136, 191
vocabularies, XML 21

W3C Recommendations 3
  CSS 162
  ITS 170
  SVG 177
  XHTML 159
  XPATH 150
  XSL-FO 148
  XSLT 148, 164
Web browsers
  CSS support 158, 163
  Internet Explorer and XSLT support 164
  SVG support 159, 183
  XHTML support 158
  XML support 158
  XSLT support 158, 164
Wireless Markup Language (WML) 4
WordProcessingML (WordML) 133
workflow, controlling 192
World Wide Web Consortium (W3C)
  Recommendations 3
  XML initiative 3

Xerces parser 52, 64
XHTML
  adoption of 161
  advantages 160
  compared with HTML 7, 159
  device-specific versions 159
  doctypes 159
  HTML Tidy tool 139
  migrating from HTML 139
  namespace 159
  overview 158
  W3C Recommendations 159
XLIFF 173
XLink 4
XMetal 6, 83, 126, 141
XML
  advantages 196–200
  advantages for technical communicators 5, 197
  advantages, general 4, 193
  and content management systems 136
  and FrameMaker 128–133
  and localization 165–176
  and Word 133–136
  as an open standard 5, 197
  authoring 6
  browser support 158
  catalog files 65
  content management 136, 187
  content management systems (CMS) 136
  disadvantages 199
  documentation source formats 5
  extensibility 198
  governmental standards 198
  in technical communication 5
  industry support 199
  localization advantages 165, 199
  localization processes 172
  markup 11–19
  metadata 4, 199
  migrating from FrameMaker 141–144
  migrating from HTML 139
  migrating from SGML 140
  migrating from Word 144
  namespaces 19–20
  origins 3
  overview 1–4
  parsing 52
  publishing through FrameMaker 129
  publishing tools 7
  return on investment 198
  reuse, advantages of 197
  schema documents 21
  semantic tagging 4
  Semantic Web, role in 199
  separation of content from formatting 7, 167
  single-source publishing 8
  standards organizations 199

technical communication tasks 193
translation, facilitating 170
validation 52
validity 18
vocabularies 21
web browsers, displaying in 158
web, using on 157
XML authoring 126
editors, desirable features 126
FrameMaker authoring 128
Microsoft Word 133
text editors 125
XML editors 126
XML editors 125
Arbortext 7, 83, 126, 136
Authentic 128
MadCap Flare 126
XMetal 6, 83, 126, 141
XMLSpy 6, 126, 144, 182
XML empty elements 13
xml lang attribute 15, 169
XML languages
Chemical Markup Language (CML) 4
DITA 66-85
DocBook 55-66
Extensible Stylesheet Language (XSL) 147
Global Information Management Metrics Exchange (GMX) 175, 205
Internationalization Tag Set (ITS) 170
Mathematical Markup Language (MathML) 4
Really Simple Syndication (RSS) 4
Resource Description Framework (RDF) 5
S1000D 5, 85-124
Scalable Vector Graphics 177
Segmentation Rules eXchange (SRX) 175
Synchronized Multimedia Integration Language (SMIL) 178
TermBase Exchange (TBX) 175
Translation Memory eXchange (TMX) 174
Wireless Markup Language (WML) 4
WordProcessingML (WordML) 133
XHTML 159
XLink 4
XML Localization Interchange File Format (XLIFF) 173
XML Schema language 39
XSL-FO 154
XSLT 148, 163
XML localization 165-176
advantages 165
automation of localization processes 168
best practices 176
character encoding 166
identifying languages 169
reuse 169
segments 174
text memory 175
tools, support of XML 175
translation units 173
XML Localization Interchange File Format (XLIFF) 173
XML names 15
XML parsing
Document Object Model (DOM) 53
MSMXL 53
Simple API for XML (SAX 53
validating parsers 52
Xerces parser 53
XML Schema language 39-52
annotations 47
attribute groups 46
complex types 43
declaring attributes 46
declaring elements 41
derived types 41
empty elements 45
example 50
frequency 44
importing declarations 48
instance documents 39
Microsoft Word support 134
mixed content 45
namespaces 40
notations 49
overview 39
redefining declarations 48
schema documents 39
simple types 41
XML documents, associating with 49
xml space attribute 15
XML syntax
attributes 14
CDATA sections 18
character references 16
child elements 13
comments 18
document elements 13
document type declarations (DTDs) 13

elements 13
end tags 13
entity references 16
mixed content 13
names 15
parent elements 13
predefined entity references 17
processing instructions 17
root elements 13
special attributes 15
start tags 13
XML declarations 13
XML4J tool 64
XMLSpy 6, 126, 144, 182
XP tool 64
XPath
data model 150
expressions 150, 153
overview 150
tree nodes 150
XSL 147
XSL Formatting Objects, see *XSL-FO*
XSL Transformations, see *XSLT*
XSL-FO
elements 155
example 156
facilities 154
formatting object (FO) processors 154
language 155
processing 154
SVG, including markup 155
XSLT
browser support 164
content management, role in 192
DocBook stylesheets 154
elements 151
example 152
facilities 149
processing 151
stylesheets 163
technical publications, uses for 149
template rules 151
XML documents, associating with 163
XPath 149
XSLT stylesheets, use of in localization 168

CPSIA information can be obtained at www.ICGtesting.com
233933LV00001B/95/P

9 780950 645988